Contents at a Glance

Table of Contents

Introduction

Windows 11 is the latest generation of Microsoft's operating system, the master program that makes a computer useful and provides support to other programs, including word processors, photo viewers, and web browsers. Much as an education equips you to read a novel or play a game, Windows 11 equips your computer to perform a wide range of activities. You can use Windows 11 and other software (or *apps*) to read or write a novel, play games or music, and stay in touch with friends and family around the world.

As Windows has evolved over the past 30 years, so have computers — the *hardware.* Today, you can buy a computer as small as a paperback book, and even such a little computer is unimaginably more powerful than (and a fraction of the cost of) computers just 10 years ago. The hardware consists of the screen, as well as optional components such as a keyboard and a mouse.

You don't need much time with a computer to conclude that there has to be an easier way to do things. At times, computers seem overly complex and inscrutable. Have you used a smartphone lately? Or a TV remote control? Why are the controls on every microwave oven different? Why does every new tool offer countless options you don't want that obscure the ones you do want? Well, we don't have the answers to those questions, but we do have step-by-step instructions for many tasks you want to perform using Windows 11.

After many years of working with computers, we find that they reward patience, curiosity, and a little methodical exploration. Seniors, in particular, know that learning never really stops and that learning new things keeps one young, at least figuratively. By the end of this book, you may be a multitasking computerist performing virtual gymnastics with Windows 11. On the other hand, if this book helps you do only one thing — use email, browse the web, or enjoy photos or music — that one useful thing may be all you need.

About This Book

Age is just a number. This book is intended for anyone getting started with Windows 11 who wants step-by-step instructions without a lot of discussion. Numerous figures with notes show you the computer screen as you progress through the steps. Reading this book is like having an experienced friend stand behind you as you use Windows 11 . . . someone who never takes control of the computer away from you.

Windows 11 is a work in progress. Microsoft updates the Windows 11 operating system from time to time to make it more secure and agreeable to the people who use it. (Chapter 3 explains how to check for updates to Windows 11.) Because the operating system is continuously updated, the screen shots you see in this book may not exactly match what you see on your screen.

Conventions Used in This Book

This book uses certain conventions to highlight important information and help you find your way around:

» Different **methods for performing steps:** In general, you can complete a step in three ways. We list the choices as follows:

- **Mouse:** If you have a mouse, follow these instructions.

- **Touchscreen:** You may be able to touch your screen to perform tasks.

- **Keyboard:** Keyboard shortcuts are often the fastest way to do something.

When you have a choice between these methods, experiment to determine which is easiest for you.

- » **Tip icons:** These point out helpful suggestions related to tasks in the step lists.

- » **Bold:** We use bold for figure references as well as for when you have to type something onscreen using the keyboard.

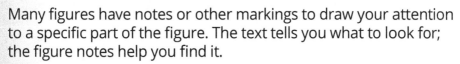

Many figures have notes or other markings to draw your attention to a specific part of the figure. The text tells you what to look for; the figure notes help you find it.

- » **Website addresses:** If you bought an e-book, website addresses are live links. In the text, website addresses look like this: www. dummies.com. See Chapter 6 for information on browsing the web.

- » **Options and buttons:** Although Windows 11 often uses lowercase in option and button names, we capitalize the text for emphasis. That way, you can find a button labeled Save Now, even though onscreen it appears as *Save now*.

How to Read This Book

You can work through this book from beginning to end or simply look at the table of contents or index to find the instructions you need to solve a problem or learn a new skill whenever you need it. The steps in each task get you where you want to go quickly, without a lot of technical explanation. In no time, you'll start picking up the skills you need to become a confident Windows 11 user.

Technology always comes with its own terms and concepts, but you don't need to learn another language to use a computer. You don't need any prior experience with computers or Windows. Step-by-step instructions guide you through specific tasks, such as accessing the news or playing a game. These steps provide just the information you need for the task at hand.

Foolish Assumptions

We assume that you have a computer and want clear, brief, step-by-step instructions on getting things done with Windows 11. We assume also that you want to know just what you need to know, just when you need to know it. This isn't Computers 101. This is Practical Windows 11. As an old friend of one of the authors said, "I don't want to make a watch; I just want to know what time it is."

How This Book Is Organized

This book is divided into four parts to help you find what you need. You can read from cover to cover or just jump to the page that interests you.

» **Part 1: Getting Started with Windows 11:** In Chapter 1, you turn on the computer and get comfortable with essential parts of Windows 11, such as the Start screen, as well as how to use a mouse, touchscreen, or keyboard. Explore features of Windows 11 apps in Chapter 2. To customize Windows 11 to work better for you, turn to Chapter 3. In Chapter 4, you create and modify user account settings, such as passwords. Discover the desktop, how to manage windows, and how to customize the desktop in Chapter 5.

» **Part 2: Windows 11 and the Web:** Use the web to stay current and keep in touch. Turn to Chapter 6 to use Edge to browse the web. Send and receive email in Chapter 7. Stay in touch with friends and family with Microsoft Teams in Chapter 8. Turn to Chapter 9 to explore a handful of apps that can help you stay in touch with friends and get to know the outside world better.

» **Part 3: Having Fun with Windows 11:** If you haven't been having any fun until now, we've failed you. Expand your tools and toys in Chapter 10 by connecting to Microsoft Store to install new apps. In Chapter 11, you enjoy photos on Windows 11 and put your own photos on the computer. If you want to listen to music and watch a video, see Chapter 12.

» **Part 4: Beyond the Basics:** In Chapter 13, you learn about the care and feeding of Windows 11, which requires a little maintenance now and then. Find out how to connect a printer and other hardware, such as a mouse and a second screen, in Chapter 14. Do you appreciate the saying "a place for everything and everything in its place"? Chapter 15 is where you find out how to organize your documents. And finally, you back up your files to insure against loss and refresh Windows 11 when it gets cranky, in Chapter 16.

Beyond the Book

In addition to what you're reading right now, this book comes with a free, access-anywhere Cheat Sheet that helps you search the web without a browser, quickly locate recently used files from the Start screen, view file extensions in File Explorer, and take a screenshot. To get this Cheat Sheet, simply go to www.dummies.com and search for "Windows 11 For Seniors For Dummies Cheat Sheet" by using the Search box.

1

Getting Started with Windows 11

IN THIS PART . . .

Master the basics.

Discover how to handle apps.

Customize Windows 11.

Manage user accounts and passwords.

Personalize the Windows desktop.

Chapter **1**

Getting Acquainted with Windows 11

Windows 11 is an *operating system* (the master program for any computer). You can use Windows 11 on a wide range of devices, from a smartphone to a big-screen TV/ entertainment system: One size fits most. You can not only use the same programs with a range of hardware but also access the documents you create (such as photos and email — files and data, to nerds) from any Windows-based computer, giving you extraordinary freedom of choice and mobility.

Although countless companies create programs you may use, Microsoft attempts to make similar functions consistent across different programs. For example, opening a document or emailing a photo to a friend involves the same steps regardless of the programs you use. You don't have to learn a different way of doing common tasks in each program. This consistency will serve you well when using Windows 11 and other new programs.

In this chapter, you start your computer and work with the *Start screen*, the dashboard for Windows 11. You explore options for using the Start screen with your *hardware* (the computer and related devices). Then you exit Windows 11 and go right back in for more.

The easiest way to get Windows 11 is preinstalled on a new computer. If your current computer runs an older version of (Windows 7, Windows 8, or Windows 10), you can upgrade to Windows 11, although older machines may lack newer functions, such as a touchscreen, and may have some hardware compatibility problems with Windows 11 as well. Also, Windows 11 may be free, or you may have to pay for it, depending on your current operating system. Visit microsoft.com for details.

Tell Your Computer What to Do

How do you get Windows 11 to do what you want it to do? You can command a computer in many ways, depending on your equipment (hardware). For example, a desktop computer has different options from a handheld phone. You may have any or all of these choices:

» Mouse

» Touchscreen

» Keyboard

Another device for controlling Windows is a touchpad, which is commonly found on a laptop keyboard. You move your finger on the touchpad to move the pointer on the screen.

If you have a computer with more than one of these devices, you might use one device exclusively or, more likely, vary your choice according to the task. Use whichever technique is easiest for you, but don't be afraid to experiment. In the next few sections, you discover the ins and outs of using all these methods of controlling Windows 11. Then you're ready to turn on your computer and use these methods.

In the steps throughout this book, *choose* or *select* refers to using a mouse, the touchscreen, or a physical keyboard. *Drag* refers to using a mouse or a finger.

Move the Mouse

A *mouse* is a soap-bar–sized device that you move across a desk with your hand. Move the mouse and note how the arrow called a *mouse pointer* moves across the computer screen. (Similar to a mouse, a *trackball* is a ball in a socket that you can rotate to move the mouse pointer.) A mouse has two or more buttons; some also have a scroll wheel between the buttons.

The following terms describe methods for using a mouse with Windows 11. In each, move the mouse first to position the pointer over a specified item before proceeding:

» **Click:** Move the onscreen arrow-shaped mouse pointer over a specified item and press and release the left mouse button: That's a click (sometimes called a left-click to distinguish it from a right-click).

» **Right-click:** Press and release the right mouse button to display available functions. Note that the word *click* by itself means use the left mouse button.

» **Drag:** Press and hold down the left mouse button, and then move the mouse pointer across the screen. When you want to move an object, you drag it. Release the mouse button to release the object.

TIP

Watch for the word *click* to indicate using a mouse button and *roll* to indicate using the mouse wheel.

Touch the Screen

A *touchscreen*, as the name says, enables you to touch the screen to tell your computer what to do. You typically use one finger or two, although touchscreens may allow you to use all ten digits. In some cases, you can also use a special pen called a *stylus* instead of your finger. Tablet computers and some smartphones have touchscreens. Touchscreens are less common on desktop or laptop computers, but that situation is changing. Not sure what type of screen you have?

When you have Windows 11 running, give the screen a gentle poke with your index finger to see what happens.

The following terms refer to ways you interact with a touchscreen:

» **Tap:** Briefly touch the screen. You *select* an object, such as a button, by tapping it.

» **Drag:** Touch and hold your finger on the screen, then move your finger across the screen. You *move* an object, such as an onscreen playing card, by dragging it.

» **Swipe:** Touch and move your finger more quickly than with drag. You can swipe your finger across the screen from any of the four sides of the screen to display options and commands. You swipe pages to move forward or back. You may see the word *flick* instead of *swipe.* Some people insist that a flick is faster or shorter than a swipe, but let's not get caught up in that.

» **Pinch and unpinch:** Touch a finger and thumb or two fingers on the screen. Move your fingers closer to each other to *pinch* and away from each other to *unpinch.* Generally, a pinch reduces the size of something on the screen or shows more content on the screen. An unpinch (an ugly word) *zooms in,* increasing the size of something onscreen to show more detail.

Watch for the words *tap, swipe,* or *pinch* to indicate using your finger. Touch actions are often called *gestures.*

See the upcoming section "View the Touch Keyboard" if your computer doesn't have a physical keyboard, as is often the case with a touchscreen.

Use a Keyboard

A typewriter-like keyboard is a traditional device for controlling a computer and is especially useful when you must enter a lot of text. Special key combinations, called *shortcut keys*, are often the quickest way to do anything (though they require some memorization).

The following keys are particularly noteworthy. No offense intended to fans of keys not noted here. Although you won't use all these keys immediately, you'll find it helpful to locate each one on your keyboard.

Press indicates that you use the keyboard (physical or virtual) for the specified key or sequence of keys (just as *click* indicates a mouse action and *tap* indicates touch). Combinations of keys are not pressed simultaneously. Instead, press and hold the first key in the specified sequence, press the second key, and then release both. (We explain exceptions to this method as necessary.)

» ⊞: Called the Windows key, this key is usually located on either side of the spacebar, which is the largest key. ⊞ works by itself, as you'll soon see, and also in combination with many other keys. Throughout the book, We specify these combinations where you might use them. There will be a quiz later. (Kidding! No quizzes.)

» **Tab:** Press the Tab key to highlight an item. Press Tab repeatedly to skip items you don't intend to select.

The keyboard can be used to select objects but is less direct than using touch or a mouse.

» **Arrow keys:** Press the arrow keys to move the cursor or selection of an object in the direction the keys point (left, right, up, or down). In some contexts, Tab and the right arrow do the same thing. Sorry to be vague, but context matters at times.

» **Enter:** In most cases, the Enter key on the keyboard chooses a selection, much as clicking or tapping does. However, you may need to use the Tab key or an arrow key to select an item before pressing the Enter key.

» **Ctrl, Alt,** and **Shift keys:** These keys are used with other keys for commands. For example, press Ctrl+C to copy selected text or an object. (That is, while pressing and holding down the Ctrl key, press the C key — no need to press Shift for an uppercase C. Then release both keys.) The Shift key is used with another key for uppercase.

» **Backspace:** As you enter text, each press of Backspace erases the character to the left of the cursor.

» **Delete:** As you enter text, each press of the Delete key erases the character to the right of the cursor. On some keyboards, this key is labeled Del.

» **Function keys:** All keys function, but Function keys are labeled F1 through F12. You don't use these much in this book, but you should locate them. Laptops often have a separate Function Lock key to turn these keys on or off.

» **Page keys:** Locate the Home, End, Page Up, and Page Down keys for future reference. Use these to move the screen, a page, or the cursor. (On some keyboards, the Home, End, Page Up, and Page Down keys work as numbers when the Num Lock key is activated.)

View the Touch Keyboard

Windows 11 can display a touch keyboard onscreen. This feature is vital for devices that have a touchscreen and no physical keyboard. With a touchscreen, the touch keyboard appears automatically when the *cursor* (a blinking vertical bar) indicates that you can enter text in a box. If the touch keyboard doesn't appear automatically, you may also see a separate box floating above or below the text box. Tap that floating box to display the keyboard. Or, if you don't see a floating keyboard, tap the keyboard icon on the taskbar to the right (it's close to the date and time). To type using the keyboard, simply tap or click a letter, number, or symbol key.

TIP

Here are the different types of touch keyboards:

» The *standard layout* (also called QWERTY) appears automatically (see **Figure 1-1).** The Enter key changes depending on the context.

» The *uppercase layout,* shown in **Figure 1-2,** appears when you tap the Shift key on the standard layout.

» The *numbers and symbols layout* appears when you tap the &123 key on the standard layout. Tap the &123 key again to return to the standard layout.

» The control keys overlay appears on five keys on the standard layout when you tap the Ctrl key. The Ctrl keys are used in common tasks, such as copying (Ctrl+C) or moving (Ctrl+X) selected text. The overlay disappears automatically after you tap one of the control keys (A, Z, X, C, or V).

» Select the Settings button (refer to **Figure 1-1**). The Setting button appears on the top left of the touch keyboard (it looks like a gear). Here, you can select the Keyboard Layout. Default is selected automatically, but you can choose Small, Split, or Traditional. You can also choose to switch to a Handwriting option, which lets you write with your finger. Windows converts your scribbles to block text. Finally, you can also select Theme and Resize. This opens the Personalization window, where you can select different keyboard and size options, shown in **Figure 1-3.**

» Finally, you can add emoticons as you type. Select the emojis button (refer to Figure 1-1) and touch or search for the emoticon you want to use.

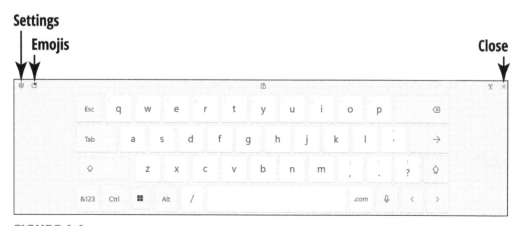

FIGURE 1-1

FIGURE 1-2

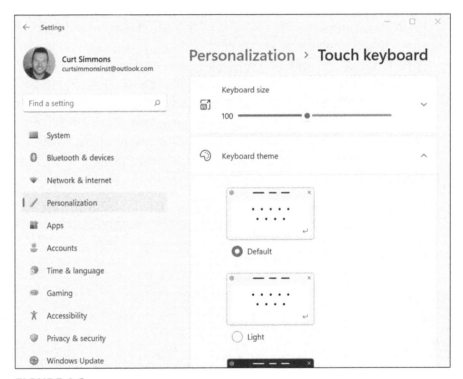

FIGURE 1-3

Turn On Your Computer

1. Push the power button briefly and release it. Every computer has a power button. (When we can no longer turn them off, the machines win.) If you have a desktop computer tower, the power button is probably on the front of the tower. Otherwise, you might have to feel around the front and sides of the screen or near the hinges of a laptop. Typically, your computer will beep, some buttons will light, and the screen may flash a logo or a message that disappears before you can read it. (Just let that go.) Soon, you will see the Windows 11 Lock screen.

2. Turn on any separate hardware (such as a monitor, speakers, or a printer), if necessary.

3. Enter your password and press Enter (or select the Submit button, the button on the right side of the password textbox). Soon the Windows 11 desktop screen appears, as shown in **Figure 1-4.**

FIGURE 1-4

If you don't see the Password text box, jiggle the mouse or press a key on the keyboard to wake up Windows 11.

TIP

The first time you turn on a new computer, a series of Windows Setup screens appears. Accept the defaults or change them appropriately and then select the button labeled Next.

TIP

If your computer doesn't have a keyboard, as is the case with many tablet computers, see the preceding section, "View the Touch Keyboard," for information on how to type onscreen.

TIP

TIP

If this is the first time that Windows 11 has started on your computer, you must create a user account, even if no one else will use the machine. See Chapter 4 for details on creating and changing user accounts.

Check Out the Start Screen

1. Start your computer — if it isn't started already — and sign in to your user account. You'll see the Windows desktop screen (refer to **Figure 1-4**).

2. Open the Start screen, as shown in **Figure 1-5.** Use one of these methods to open the Start screen:

   • **Mouse:** Click the Start button (you'll find it on the taskbar, which is the strip across the bottom of the screen). The Start button is the leftmost icon on the taskbar; it looks like a four-pane window.

   • **Touchscreen:** Tap the Start button.

   • **Keyboard:** Press the ⊞ key.

3. Examine the Start screen and note the icons. These icons represent available *apps* (short for application programs, an older term for programs or software). By clicking or tapping an app icon, you can open an app. (See Chapter 2 for information on using individual apps.)

4. You can take a look at more apps by clicking the Next Page button (refer to **Figure 1-5**). Also, you can see an alphabetical list of all the apps that are installed on your computer, as shown in **Figure 1-6,** by clicking the All Apps button (refer to **Figure 1-5**). By clicking or tapping an app in this list, you can open an app. Click the Back button to go back to the main Start Screen page.

TIP

Under Recommended, you see recent files that you've opened. Click More to see a complete list. This is a handy way to find a file you recently used.

Apps　　　　　　　　　　　　　　　**All apps button**

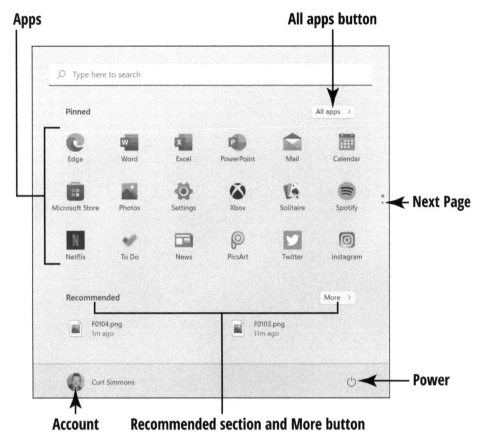

← **Next Page**

← **Power**

Account　　**Recommended section and More button**

FIGURE 1-5

TIP

While the Start screen is open, you can type the name of an app to open it. For example, to open the Weather app, type **weather.** This automatically changes the Start screen to the Search screen, which you can also open directly by clicking the Search icon on the taskbar. Select the Weather app listing in the search results to open the Weather app.

TIP

Rather than scroll through the alphabetical list to find an app, you can select a letter in the list and then select a letter in the pop-up list of letters that appears (refer to **Figure 1-7**). For example, to quickly get to the Weather app, select any letter and then select the W on the pop-up list.

FIGURE 1-6

FIGURE 1-7

5. Click or tap your image (or image and name) on the Start screen. As shown in **Figure 1-8,** you see a pop-up menu with commands for changing account settings (see Chapter 3), locking your screen (see "Start Again on the Lock Screen," later in this chapter), and signing out in a favor of another person who shares your computer (see Chapter 4). Who would think that clicking your name would open a pop-up menu?

6. Click or tap Power on the Start menu. As shown in **Figure 1-9,** you see a pop-up menu with commands for putting your computer to sleep, shutting down your computer, and restarting your computer. Later in this chapter, "Shut Down Your Computer" looks into the options on this pop-up menu.

7. Click or tap the Start button. Doing any of these actions when the Start screen is open closes the Start screen. You can also close the Start screen by clicking anywhere on the desktop when the Start screen is open.

FIGURE 1-8

FIGURE 1-9

Shut Down Your Computer

1. When you have finished using your computer for a while, you may want to shut down Windows 11. Begin by displaying the Start screen:

   - **Mouse:** Click the Start button on the taskbar. The Start button is the leftmost icon and looks like a four-pane window.

   - **Touchscreen:** Tap the Start button.

   - **Keyboard:** Press the ⊞ key.

2. Click or tap Power on the Start menu (refer to **Figure 1-9**). The Power button is on the bottom right side of the Start menu and looks like a circle with a line through the top.

3. Available options appear in a pop-up box. Some or all of the following options appear:

   - **Sleep:** This option reduces the computer's power consumption without exiting Windows 11 or closing apps (some computers don't offer this option). As a result, when you wake the computer by moving the mouse or touching the screen or the keyboard, everything is exactly as you left it: Apps and documents are open, if they were open before Sleep.

- **Shut Down:** This option exits Windows 11 and saves power by turning the computer off. In exiting Windows 11, Shut Down closes any apps that are currently running.

- **Restart:** Temporarily shuts down Windows 11 and turns it on again. Use Restart when Windows 11 asks you to or when Windows 11 is misbehaving.

TIP

You can also shut down your computer by using the Start button. Move the pointer over the Start button and right-click. A pop-up menu appears. Select Shut Down or Sign Out on the pop-up menu, and then select an option (Sleep, Shut Down, or Restart).

4. Choose Shut Down to turn off the computer.

TIP

On most computers, pressing the power switch also shuts down the computer, though you should use the Shut Down option in Windows 11 rather than press the power switch button. On a laptop, closing the lid may shut down the laptop or put it in sleep mode.

Start Again on the Lock Screen

1. Turn on your computer. Every time you turn on your computer, the Lock screen appears. As shown in **Figure 1-10,** the Lock screen displays the time, day, and date along with a photo. (You discover how to change this photo in Chapter 3.)

2. Dismiss the Lock screen with one of these methods:

- **Mouse:** Click anywhere, roll the wheel toward you, or drag the entire screen up.

- **Touchscreen:** Drag the entire screen up.

- **Keyboard:** Press any key.

3. If you don't use a password or Windows Hello to sign in, wait briefly for the Start screen to appear. If you use a password, enter it with a physical or touch keyboard. Then press Enter or select the arrow next to the password box to display the Windows desktop screen.

4. Take a break before reading Chapter 2.

FIGURE 1-10

TIP

When you take a long break from your PC, it automatically goes to sleep. Windows 11 puts your computer to sleep to save battery life. To decide for yourself how much time passes before your computer goes to sleep, select the Search icon on the taskbar and type **lock screen** in the Search box. Choose Lock Screen Settings in the search results. On the Personalization > Lock Screen window, scroll down the screen to the Screen Timeout and select it. You come to the Power & Sleep screen. Choose a setting from the Sleep menu. For example, choose 15 minutes to put your computer to sleep after 15 minutes of inactivity.

Chapter **2**

Using the Start Screen, Apps, and Widgets

The Windows 11 *Start screen* appears when you select the Start button on the taskbar or you press the ■ key. It provides access to just about everything you do with Windows 11.

The Start screen is home to numerous programs, or *apps* (short for applications). An app performs a function or displays information. For example, the Weather app offers weather reports (surprise!) and the Maps app helps you map a route between two places. Apps can be simple or complex.

Apps appear on the Start screen as *icons.* An icon displays the app's name and symbol for the app. An open app typically covers the screen, hiding other apps that are open. However, some apps can be displayed side-by-side with a function called *snap.*

Windows 11 also provides easy access to a group of mini apps called *widgets.* Widgets are graphical apps that give you fast at-a-glance information about news, sports, weather, stocks, and so forth. You choose what widgets you want to see and quickly view them using the Widgets icon on the taskbar.

In this chapter, you open, close, and use some of the apps that come with Windows 11. You also switch between apps and the Start screen, and switch directly between two apps. You find out how to search for apps not shown on the Start screen. You discover how to organize the Start screen by rearranging and pinning apps. Finally, you explore how to add and remove widgets.

See Chapter 10 for information on getting new apps from the Micro-soft Store.

Although some steps are specific to one app, most of the steps you follow in this chapter can be repeated in any app.

TIP

Open Windows 11 Apps

1. Select the Start button taskbar, as shown in **Figure 2-1.** Selecting this button opens the Start screen, also shown in **Figure 2-1.**

   You can also open the Start screen by pressing the ⊞ key.

2. On the Start screen, you see a group of apps. These apps are *pinned* to the Start Screen, meaning that they always appear (you can change what's pinned, and you find out how in "Pin and Unpin Apps on the Start Screen," later this chapter). Click the All Apps button to see an alphabetical listing of all apps available on your computer. Select All Apps and then locate the Weather app and select it. The Weather app opens to the Forecast screen, as shown in **Figure 2-2.** It shows the current temperature and weather forecast for your default location. Select the Show Options button (see **Figure 2-2**). As do most apps, Weather has a Show Options button in the upper-left corner. Select this button to expand the app bar and see the names of options on the app bar. Select the button again to collapse options on the app bar.

The first time you open the Weather app, a Welcome screen appears. It asks whether you want to show temperatures in Fahrenheit or Celsius and what your location is. By location, the Weather app wants to know where you live, or, if you're a vagabond, where you spend the majority of your time. Choose Detect My Location or enter a city or town name in the Search box, and from the menu that appears as you type, select the name of the town or city that you call home. You can always change these settings by selecting the Settings button in the app bar and choosing options on the Settings screen. (See "Change App Settings," later in this chapter.)

3. Switch back to the Start screen using one of these methods:

- Tap or click the Start button.

- Press the ⊞ key.

Focus on the method you think is easiest. However, keep in mind that alternative methods of controlling your computer are always available.

4. Switch back to the Weather app by selecting its title bar with the mouse or your finger (it's now behind the Start screen). The Weather app reappears and the Start screen disappears.

5. Switch back to the Start screen.

6. Select the Microsoft Edge icon (it's probably pinned in your app list on the Start screen). The Edge screen appears, as shown in **Figure 2-3.**

If the Edge icon doesn't appear on your Start screen, select the All Apps button and locate it in the alphabetical list.

Edge is a browser, which is an Internet application for exploring the web. Chapter 6 covers Edge in detail.

7. Scroll downward to see all that is on the web page you are visiting. Use these techniques to scroll:

- **Mouse:** Drag the scroll box on the right side of the screen up or down. If your mouse has a wheel, you can also turn the mouse wheel to scroll.

- **Touchscreen:** Swipe the screen up or down.

8. Close Edge and the Weather app. Use one of these methods:

- **Mouse:** Click the Close button (the X) in the upper-right corner of the app's window.

- **Touchscreen:** Tap the Close button (the X) in the upper-right corner of the app's window.

- **Keyboard:** Press Alt+F4.

Start screen

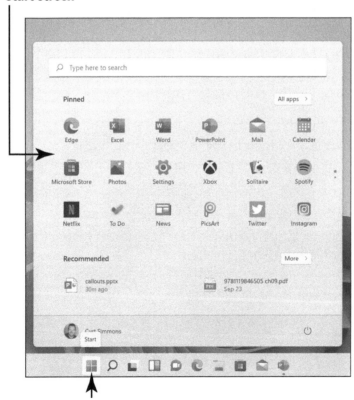

Start button

FIGURE 2-1

App bar

Show Options

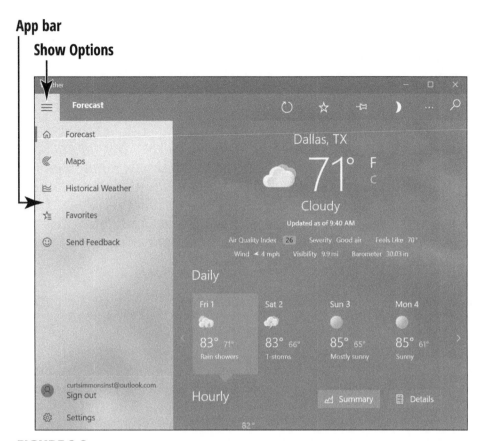

FIGURE 2-2

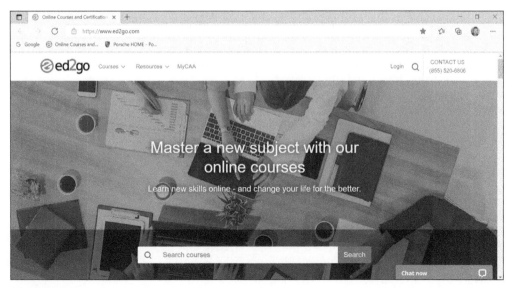

FIGURE 2-3

Chapter 5 explains how you can open a second desktop on the screen and in so doing keep some of your open apps on one screen and some of your open apps on another. When you're running many apps, opening a second desktop is a great way to be able to switch quickly from one app to another.

Open apps appear as icons on the taskbar, but when you close an app, its icon disappears from the taskbar. Some icons appear permanently on the taskbar. For example, the File Explorer icon is always on the taskbar regardless of whether File Explorer is running. Chapter 5 explains how you can pin your favorite apps to the taskbar. Pinning an app to the taskbar places an icon there so that you can open an app quickly.

Press Alt+Tab and continue to hold down the Alt key after you press Tab. A window showing thumbnails of all open apps appears onscreen. While holding down the Alt key, press the left- or right-arrow key to move between the open apps to select one; then release the Alt key. The Weather app appears onscreen. Pressing Alt+Tab is yet another way to switch between open applications.

Chapter 5 demonstrates techniques for moving windows onscreen, changing the size of windows, and snapping windows to the side of the screen.

To close an app that is frozen (an app that is unresponsive), press Ctrl+Shift+Esc. The Task Manager opens. It lists all apps that are currently running. Select the app you want to close and then select the End Task button.

Use the App Bar

1. From the Start screen, open the Weather app if it isn't already open.

2. The *app bar* contains functions specific to the current app. Display the app bar by selecting the Show Options button. (Refer to **Figure 2-2**.)

The app bar may appear across the top of the screen, the left side of the screen, or in both locations.

3. In the Weather app, the app bar leads you to different functions. Select Maps, for example, to see a weather map of the area where you live, shown in **Figure 2-4.**

4. Display the app bar in Weather again. Then select Historical Weather. A screen appears showing you weather trends over the past 12 months.

5. Display the app bar again and take notice of the Home icon. Wherever you travel in an app, you can return to the app *home screen* by selecting this icon.

FIGURE 2-4

In most apps, you can select the Back button (a left-pointing arrow) to return to the previous screen you viewed. The Back button is found in the upper-left corner of most screens.

Add a Location in Weather

1. From the Start screen, open the Weather app if it isn't already open. With the Weather app on the screen, select the Show Options button to expand the app bar and see the option names (refer to **Figure 2-2**).

2. Select the Favorites button on the app bar. The Favorites screen appears, as shown in **Figure 2-5.** Your screen will show a different Launch Location.

3. Select the Add to Favorites icon, which looks like a plus sign in a square. The Add to Favorites screen appears.

4. Type a location name, such as a city, in the box under Add to Favorites, as shown in **Figure 2-6.** As you type, matching location names appear below the box. If you see the location you want, select that name to add an icon for that location to the Places screen. No need to click the Add button, unless your location does not appear automatically.

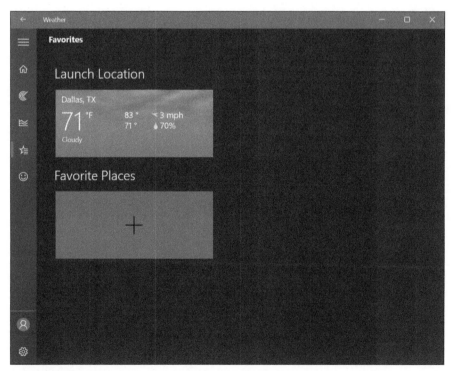

FIGURE 2-5

You can add other locations by repeating Steps 3 and 4.

5. Select the icon for the location you added. The Weather app displays full information for the location you selected.

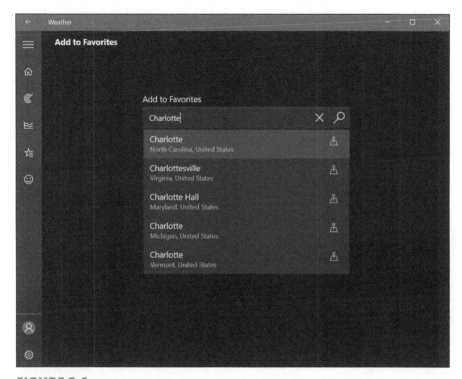

FIGURE 2-6

You can switch among multiple locations by using the Favorites button on the app bar.

Select Favorites in the app bar to see a screen with weather locations you chose (refer to **Figure 2-6**). From there, you can select a weather location to get reports and predictions about weather in a different city or place.

Change App Settings

1. On the Start screen, select the Weather icon if the Weather app isn't open already.

2. In the Weather app, select the Settings button on the app bar. The Settings screen appears, as shown in **Figure 2-7.** Settings is the bottommost option on the app bar. If you have trouble locating it, select the Show Options button, the topmost button on the app bar, to see the buttons on the app bar.

TIP

It may be hard to remember whether you need the Settings panel or the app bar to do something. In general, functions on the app bar are used more frequently than those on the Settings panel. When in doubt, guess — that's how discoveries are made.

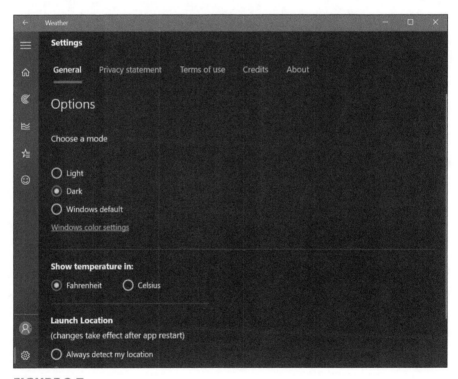

FIGURE 2-7

3. Choose whether to show air temperatures in Fahrenheit or Celsius.

4. Choose a Launch Location option. The Always Detect My Location option tells the Weather app to note by way of your Internet connection where you are currently and give the weather report for that place. Select Default Location and enter the name of the place where you live if you want to receive weather forecasts for that place.

TIP

Select Privacy Statement in the Settings window if you're interested in how Microsoft collects data about you when you use Weather and other apps made by Microsoft. You see a page clotted with legalese and other obfuscations.

Search for an App

1. Open the Start screen. A Search box appears at the top, as shown in **Figure 2-8.** If you select the Search box to type a search, the screen automatically changes to the Search screen, shown in **Figure 2-9.** More easily, you can just select the Search icon on the taskbar to search for an app (or anything else!).

2. In the Search box, type **calc** (the first four letters of the word *calculator*). The Search window shows you the results of your search, as shown in **Figure 2-10.** You can search for just about anything from the Search box.

TIP

Select a filter option at the top of the Search panel (refer to **Figure 2-11**) to redirect your search. For example, select Apps to search only for apps, Documents to search for files on your computer; select Web to search the web. You can select More to open a drop-down menu and search for any number of things, including folders, music files, photos, and videos.

3. Type **ulator** to finish typing the word *calculator*. The Search panel lists only items with the complete word *calculator* in them, including the Calculator app.

Search box

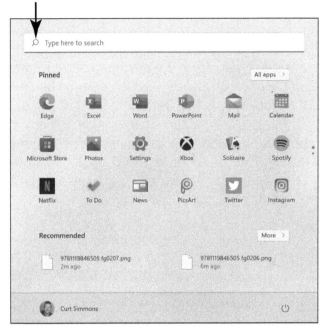

FIGURE 2-8

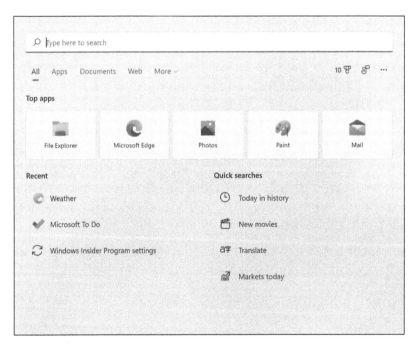

FIGURE 2-9

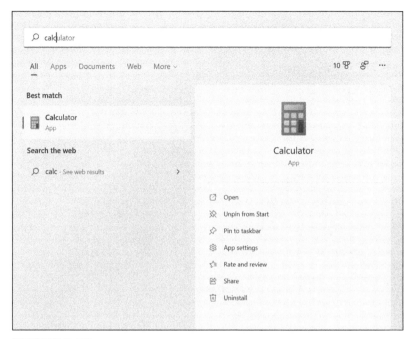

FIGURE 2-10

Search filter options

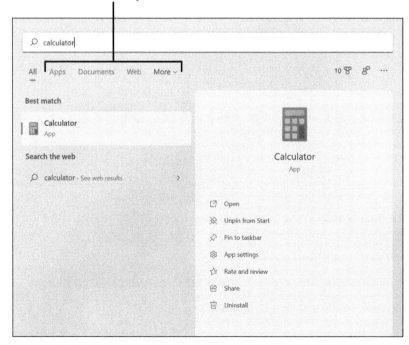

FIGURE 2-11

4. Select the Calculator app to open it on the desktop, as shown in **Figure 2-12.** To perform a calculation, select the buttons on the screen or use a keyboard.

5. Close the Calculator app by selecting the Close button.

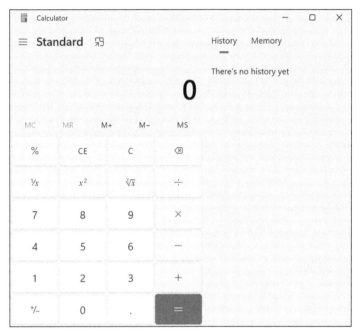

FIGURE 2-12

Pin and Unpin Apps on the Start Screen

1. Open the Start screen. Your computer has numerous apps already pinned to the Start Screen for easy and fast access. However, you may not want some of these default apps pinned, and you may want some apps pinned that are not. You can easily change the pinned apps any time. Select the All Apps button and locate the Calculator button in the alphabetical list.

2. Display the Calculator app's context menu using one of these methods:

   - **Mouse:** Right-click Calculator icon.

   - **Touchscreen:** Touch and hold the Calculator icon until the context menu appears, and then release your finger.

3. On the context menu, choose Pin to Start, shown in **Figure 2-13.**

4. Select the Back button to return to the Start screen's main window. The Calculator app is now pinned to your Start screen and appears in the Pinned group.

5. To unpin the Calculator app from the Start screen, display its context menu again and choose Unpin from Start.

FIGURE 2-13

TIP

When you unpin an app from the Start screen, it simply removes the app's icon from the Pinned group. The app is still safe and sound on your computer. You can pin and unpin any apps you want at any time so that you see only pinned apps you frequently use.

Arrange Apps on the Start Screen

1. You can rearrange icons on the Start screen by dragging them to different places. On the Start screen, drag the Calculator icon to a different location. As you move the icon, other icons move out of the way, like a game of Dodge Icon.

2. To drag an app, select the app's icon with your left mouse key and hold down the key. Then just drag the icon to a new location in your Pinned apps. If you're using a touchscreen, just press and hold the app's icon with your finger and drag.

3. Continue this process until you have organized the app icons in a way that is helpful to you. For example, you may organize the apps in a list starting with the most frequently used to the least frequently used, or you could even place them in alphabetical order. You can rearrange the pinned apps at any time by repeating these steps.

Do you have one app you want first in the list? Just display the app icon's context menu and choose Move to Top.

TIP

Uninstall an App

1. You can download apps from the Microsoft Store (see Chapter 10), but there may come a time when you want to uninstall an app. For example, say you download a game but don't find it particularly fun. You no longer want the app on your computer. Good news! You can easily uninstall it. Open the Start screen and select the All Apps button.

2. Locate the app you want to uninstall in the alphabetical list.

3. Display the app icon's context menu using one of these methods:

- **Mouse:** Right-click the app's icon.

- **Touchscreen:** Touch and hold the app's icon until the context menu appears, and then release your finger.

4. Choose Uninstall on the context menu, shown in **Figure 2-14.** A dialog box appears telling you the app will be removed. Select the Uninstall button on the box. The app is then removed from your computer.

FIGURE 2-14

TIP

After you uninstall an app from your computer, it's gone. The only way to get it back is to redownload it from the Microsoft Store.

Use the Widgets Panel

1. On the taskbar, select the Widgets icon to open the Widgets panel, shown in **Figure 2-15.**

Widgets icon

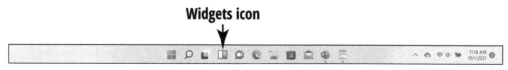

FIGURE 2-15

2. The Widgets panel has three main sections:

- **Search:** The Search box gives you easy way to the Web. Search results appear in Microsoft Edge when you press Enter (see Chapter 6 for more information about Microsoft Edge). See **Figure 2-16.**

- **Widgets:** This section shows you the widgets currently installed.

- **News:** If you scroll down the panel, shown in **Figure 2-17,** you see top news stories of the day. You can select a news story to view it, and it will open in Microsoft Edge.

3. To add a new widget, select the Add Widgets button (see **Figure 2-16**). In the Widget Settings window, select the plus sign next to the widget you want to add (see **Figure 2-18**). For example, try adding the Traffic widget by selecting the plus sign next to its name. You can add multiple widgets at the same time. When you're done, close the window by clicking the Close button.

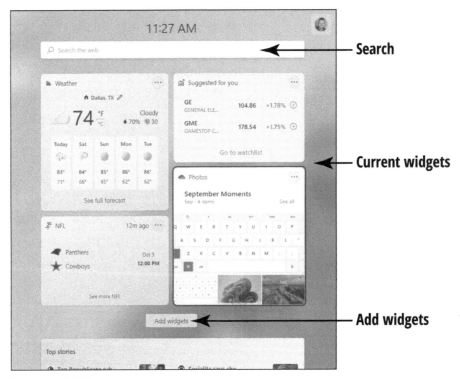

FIGURE 2-16

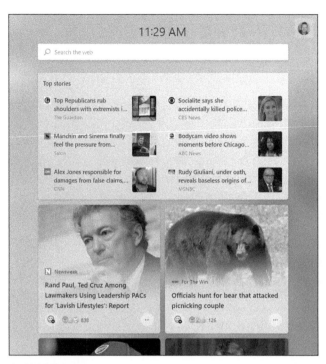

FIGURE 2-17

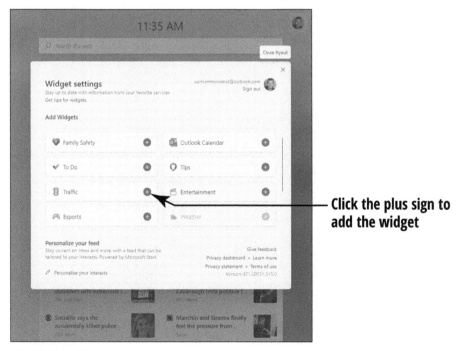

Click the plus sign to add the widget

FIGURE 2-18

Remove and Reorganize Widgets

1. Open the Widgets panel by clicking the Widgets icon on the taskbar.

2. You can drag the widgets around to display them in a different order on the Widgets panel. There's no right or wrong way to organize the widgets, so just organize them in a way that's best for you. To drag a widget to a new location on the panel, hover your mouse on the widget until you see a hand icon. For touchscreens, hold your finger on the widget. Then just drag the widget to a new location. The other widgets move out of the way, as shown in **Figure 2-19.**

3. To remove a widget, select the See Options button on the upper right corner of the widget. On the menu that appears, choose Remove Widget, shown in **Figure 2-20.**

TIP

When you remove a widget, it's just removed from your widget panel (it's not deleted). You can always add it again later using the Add Widgets button.

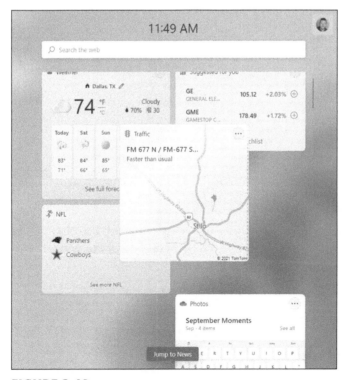

FIGURE 2-19

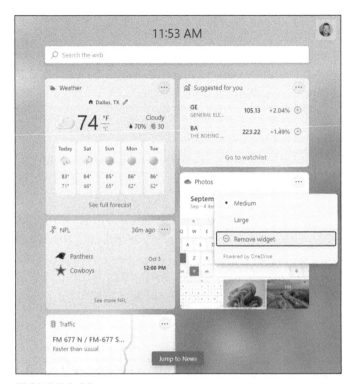

FIGURE 2-20

Customize a Widget

1. Open the Widgets panel by clicking the Widgets icon on the taskbar.

2. You can change the size of widget. Select the See Options menu located in the upper right corner of the widget. You can choose from Small, Medium, or Large. In **Figure 2-21,** all the widgets are set to Small.

3. To change the size of the widget, just select the size you want from the See Options menu. Widgets do not have to be same size. For example, in **Figure 2-22,** the Weather widget is set to Large and the Traffic widget is set to Medium. The other widgets are set to Small.

TIP

Larger widgets typically display more information. Consider making your most important widgets larger and keep the less important widgets smaller.

4. You can also customize the data you see on each widget. The customization options vary based on the widget and what the widget does. For example, add the Traffic widget to your widgets (if it's not already added). Then select See Options and choose Customize Widget, as shown in Figure **2-23.**

5. On the customization window, you can specify a location, such as a city and state, or you can choose to detect your location, shown in Figure **2-24.** The second option is helpful if you travel frequently; that way, the widget is always accurate wherever you're located. Make your selection and choose Save.

TIP

Remember, widgets are designed to give you quick information. The idea is that you open the Widgets panel and immediately see information that is important to you. Keep this thought in mind as you choose, organize, and customize your widgets.

FIGURE 2-21

FIGURE 2-22

FIGURE 2-23

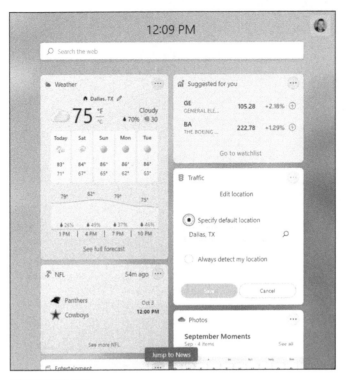

FIGURE 2-24

Chapter **3**

Adjusting Windows 11 Settings

O ut of the box, Windows 11 is showy and colorful. If you don't like that look, however, you can change the photos and colors you see on the screen. Adjusting Windows 11 settings can also make Windows 11 easier and more fun to use. When you're ready, you can dive in and make Windows 11 yours.

In this chapter, you personalize the Lock screen. You see this screen many times a day, so it should please you. You also choose a picture to identify your account on the Start menu. This chapter also explains how to make your screen easier to see and enable features such as Narrator, which reads aloud content from the screen. Finally, you discover how to customize the Start menu and decide which notifications pop up in the lower-right corner of your screen.

Many people leave Windows 11 largely as they found it. Some love to tweak, tinker, and tune. How far you go in personalizing Windows 11 is up to you — it's your computer, after all.

TIP

See Chapter 4 for information on changing passwords and other User settings.

Access the Settings Screen

1. Select the Search button on the taskbar and search for Settings.

2. Choose the Settings app in the search results, as shown in **Figure 3-1**.

TIP

Turn to Chapter 1 if you need help opening an app on the Start menu.

3. The Settings screen appears, as shown in **Figure 3-2**. The Settings screen is the starting point for changing the settings here, there, and everywhere on your computer.

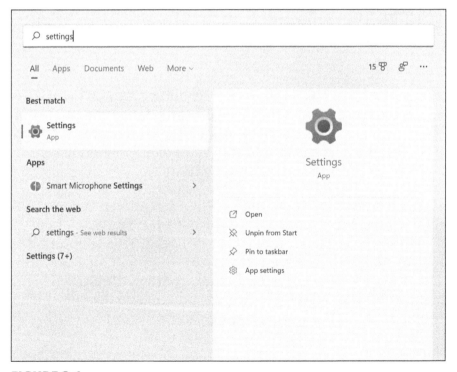

FIGURE 3-1

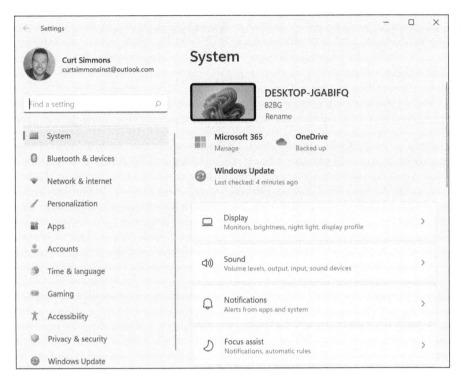

FIGURE 3-2

4. One at a time, select each icon in the Settings screen, starting with System, to see the options available. Select the Back button (the left-pointing arrow, located in the upper-left corner of the screen) to return to the Settings screen.

5. On the Settings screen, enter the word **printer** in the Search box (located at the top of the screen; refer to **Figure 3-2**). As soon as you enter the word, the Settings screen shows you a list of settings that pertain to printers. You can select a setting on the list to open the dialog box where the setting is located.

TIP

Windows 11 offers many different settings. Locating them can be a chore. Often the easiest way to find a setting is to enter its name in the Search box on the Settings screen. Also, you can search for any setting using the Search icon on the taskbar.

Personalize the Lock Screen

1. On the Settings screen, select Personalization.

2. On the Personalization screen, scroll down and select the Lock Screen option on the right side. The Lock Screen Preview screen opens, as shown in **Figure 3-3.** On the Personalize Your Lock Screen drop-down menu, you see three options, as shown in **Figure 3-3.**

   - **Windows Spotlight:** This option displays an image that Microsoft chooses. The image will change frequently, giving you a nice variety of beautiful photos (typically nature shots).

   - **Picture:** This option allows you to choose a photo from your collection that is currently on your PC (such as in your Pictures folder).

   - **Slideshow:** This option allows you to rotate through a collection of photos that will appear on the Lock screen as a slideshow. If you choose this option, all the images need to be in the same folder on your computer.

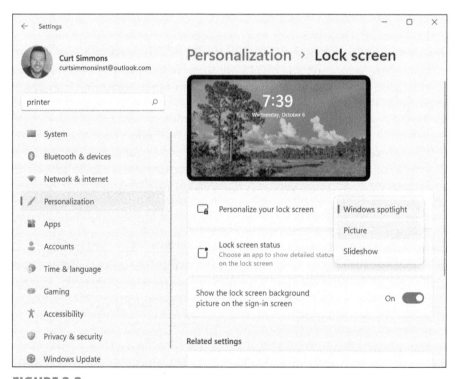

FIGURE 3-3

Choose Picture to select one of your images.

TIP

The Lock screen is the first screen you see when you start Windows 11.

3. You can select from recent images that appear after you select the Picture option. You can also click the Browse Photos button to search for a specific photo, as shown in **Figure 3-4.**

TIP

You can use the Browse button to select one of your own photos from the Pictures folder. See Chapter 15 for information on adding photos to this folder and its subfolders.

4. After you select a picture, it will display on the Lock screen. You repeat these steps at any time to choose a different picture, or you can choose Slideshow or Windows Spotlight.

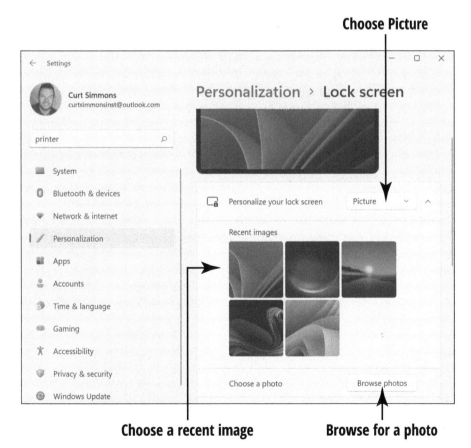

Choose Picture

Choose a recent image   **Browse for a photo**

FIGURE 3-4

TIP

You can lock your computer anytime by pressing ⊞+L.

Choose an Account Picture

1. On the Settings screen, select Accounts on the left panel. See "Access the Settings Screen," earlier in this chapter, if you need help opening the Settings screen. Your current Account picture appears in the Your Info section on the right, as shown in **Figure 3-5.** (If you haven't selected a picture yet, your picture may just be an outline.)

TIP

Your account picture appears on the password screen and the Start screen as well as other locations in Windows 11 where your account is noted.

2. To choose one of your photos, select the Browse Files button. The contents of your Pictures folder appears in the Open dialog box. From here, you can select the File Explorer navigation pane to access any folder on your computer or select a subfolder to open. When you find the picture you want, select it and then select the Choose Picture button. Or select Cancel to return to the previous screen without changing your account picture.

TIP

See Chapter 11 for information about adding photos to your Pictures folder.

3. If you have a built-in or attached camera (called a *webcam*), select the Open Camera button to take a photo. The Camera app opens with a preview of what your camera sees. See Chapter 11 for information on taking pictures with a webcam.

4. Return to the Start screen to see your new account picture. To return to the Start screen, click or tap the Start button or press ⊞ on your keyboard.

TIP

The easiest way to access the Accounts screen and change your account picture is through your account name. On the Start screen, right-click your account picture and select Change Account Settings, as shown in **Figure 3-6.** Voilà! The Accounts screen appears, with Your Info selected.

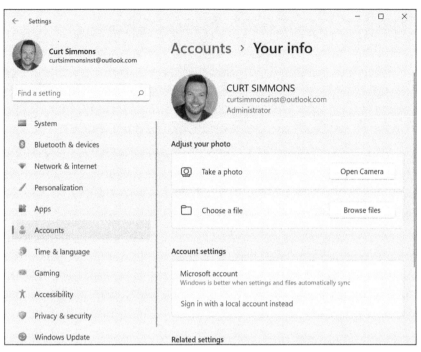

FIGURE 3-5

FIGURE 3-6

Check for Important Updates

1. To check for updates to Windows 11, select Windows Update on the left panel of the Settings screen. See "Access the Settings Screen," earlier in this chapter, if you need help getting to the PC Settings screen. The Windows Update screen appears, shown in **Figure 3-7.**

2. The Windows Update screen informs you when Windows 11 last checked for updates and whether any were found. Select the Check for Updates button to find out whether any updates are available and update Windows 11 on your computer.

3. For some updates to be installed, your computer has to shut down and restart. In cases like this, you see the Restart Now button on the Windows Update screen. You can select the Restart Now button to install the updates right away or just wait until the next time you shut down and restart on your own to install the updates.

Another way to handle updates is to simply leave your computer on overnight. Windows 11 performs the update as you sleep.

TIP

4. You may see a message that one or more updates will be downloaded and installed automatically. You do not have to do anything to install these updates — the update process is automatic. Return to the desktop to let Windows 11 manage updates automatically.

You can select the Advanced Options link on the Windows Update screen to choose how Windows 11 installs updates. Select the View Update History button and get a history of updates that Windows 11 has made to your computer.

TIP

Installing an update seldom takes more than a few minutes. You can use your system during the update process.

TIP

For information on other updates and maintaining Windows 11, see Chapter 13.

TIP

FIGURE 3-7

Make Windows 11 Easier to Use

1. On the Settings screen, select Accessibility. (See "Access the Settings Screen" for help.) The Accessibility screen, shown in **Figure 3-8,** appears. This screen offers many settings to make your computer, mouse, and monitor easier to use. These settings are organized into three categories: Vision, Hearing, and Interaction. (You need to scroll down the window to see all the settings.)

TIP

Here's a quick way to open the Ease of Access screen: Press ⊞+U (for *usability*).

2. Under Vision, choose the Text Size settings (refer to **Figure 3-8**). These settings make the screen easier to view and read by making the text larger or smaller.

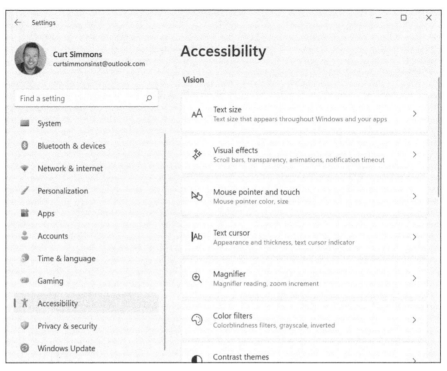

FIGURE 3-8

3. Also under Vision, choose the Mouse Pointer and Touch settings to change the size and color of the mouse pointer. As shown in **Figure 3-9,** Windows 11 offers mouse pointers of different sizes and colors. Choose the combination that tickles your fancy.

4. Also under Vision, choose Text Cursor settings to change the color and thickness of the text cursor. The *text cursor* is the vertical line that appears onscreen to show you where text will appear when you begin to type.

5. Again under Vision, choose Magnifier settings to enlarge what is on your computer screen. These settings are for people who have difficulty seeing. When Magnifier is turned on, the Magnifier toolbar appears onscreen, as shown in **Figure 3-10.** (Click or tap the magnifying glass icon on the taskbar to see this toolbar, if necessary.) Click or tap the plus sign on the Magnifier toolbar to zoom a portion of the screen for easier viewing. Click or tap the minus sign to see the screen at normal size.

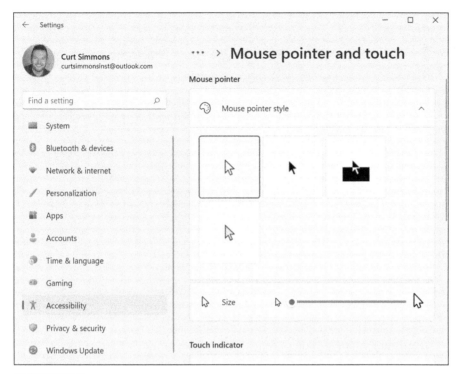

FIGURE 3-9

TIP

Without turning the Magnifier switch on, you can magnify the screen. Press ⊞+plus key to zoom in using Magnifier. Press ⊞+minus key to zoom out using Magnifier.

6. Under Vision, choose a Color Filters setting if you are color blind. If you know which type of color blindness you have — deuteranopia, protanopia, tritanopia — you can choose its name on the Color Filters menu to make seeing items onscreen easier.

7. Under Vision, choose Contrast Themes settings to alter the screen in a way that might make seeing text easier. These settings are also for people who have difficulty seeing. After you choose a theme, you can choose a color for text, hyperlinks, disabled text, selected text, button text, and backgrounds. Click Apply after you make your choices. Choose None and click Apply if you want to return to the default Windows 11 contrast settings.

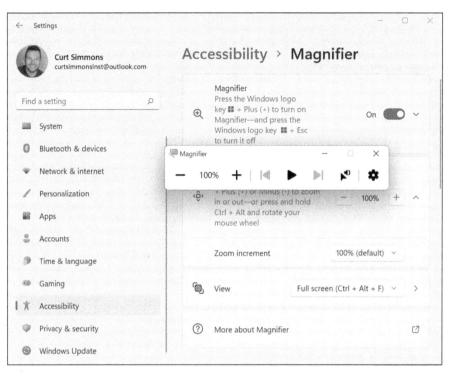

FIGURE 3-10

8. Under Vision, choose Narrator settings to have content on the screen read aloud. These settings are for people who have difficulty reading. After turning on the Narrator switch, you choose voice, speed, and pitch settings to determine what the narrator's voice sounds like.

9. Under Hearing, choose Audio Settings to increase the default volume and control whether you hear stereo or mono sound on your computer. If you have poor-quality speakers, turning the Mono Audio option to On can improve the sound.

10. Also under Hearing, choose Captions settings to control how subtitles appear in audio and video presentations. Choose Font settings to describe what you want the text in the captions to look like. Choose Background and Window settings to describe what the box where the captions appear looks like. The Preview box shows what your choices mean in real terms.

11. Under Interaction, choose Speech settings to dictate text instead of typing it and control your computer and device with voice commands.

12. Also under Interaction, choose Keyboard settings to type without using a physical keyboard. This onscreen keyboard uses the layout of a conventional keyboard. However, most people find that the standard Windows 11 virtual keyboard is more flexible. See Chapter 1 for information on the virtual keyboard layouts.

13. Again under Interaction, choose Mouse settings if for some strange reason you want to control the mouse with the keys on the numeric keypad.

14. Finally, under Interaction, choose Eye control. This feature allows you to use an eye-tracking device to control the screen by using your eyes. You have to have a tracking device for this to work. You can select the link on this screen to get more information about setting up and using an eye-tracking device.

Customize the Start Menu

1. On the Settings screen, select Personalization. See "Access the Settings Screen," earlier in this chapter, if you need help opening the Settings screen.

2. On the Personalization screen, select Start. The Personalization > Start screen opens, as shown in **Figure 3-11.** This screen offers a few ways to customize the Start screen.

3. You can turn the following items on or off (refer to **Figure 3-11**):

   - **Show Recently Added Apps:** Places the names of apps you recently acquired on the Start menu.

   - **Show Most Used Apps:** Places the names of programs you use most often on the Start menu. The names of these programs appear under the Recommended heading.

- **Show Recently Opened Items in Jump Lists on Start or the Taskbar and in File Explorer:** Allows you to quickly open folders, files, and windows from the Start menu, taskbar, and File Explorer screen. When you move the pointer over an app or program on the Start menu, an arrow appears; select the arrow to see a menu of items you can open. When you move the pointer over an icon on the taskbar, thumbnail windows appear so that you can choose which window to open. In File Explorer, the items are listed in the Quick Access section of the navigation pane.

- **Folders:** Allows you to place specific folders on the Start menu. These folders appear next to the Power button after you place them there. For example, you might want to add your Documents or Pictures folder to the Start menu for fast and easy access.

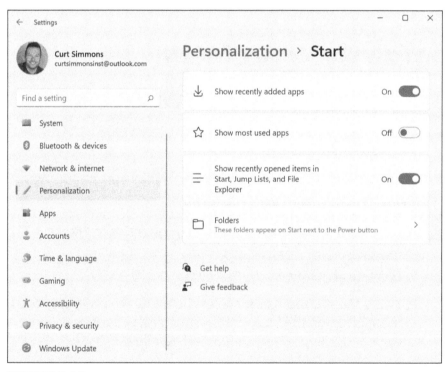

FIGURE 3-11

Handle Notifications

1. *Notifications* are the sometimes pesky little message boxes that appear in the lower-right corner of the screen when you least expect them. You can decide for yourself whether notifications appear and which notifications you want to see. On the Settings screen, select System. See "Access the Settings Screen," earlier in this chapter, if you need help opening the Settings screen.

2. Select Notifications to see the screen shown in **Figure 3-12.**

3. Under Notifications in the Notifications from Apps and Other Senders list, turn options on or off to tell Windows whether you want notifications to appear on your screen.

4. Select the Close button (the *X*) to close the Settings window.

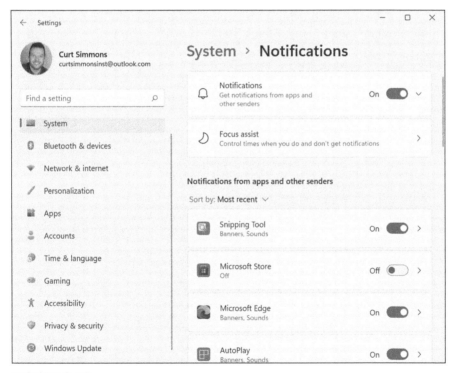

FIGURE 3-12

Chapter **4**

Working with User Accounts

Windows 11 seeks an Internet connection automatically from the moment you start it. More often than not, you connect to the Internet using a wireless or *Wi-Fi* connection. For this reason, if you start a laptop or tablet in a coffee shop or library, you may see a notification that one or more network connections are available. That's convenient.

A computer without an Internet connection is an island, if not a paperweight. Connecting to a network, however, opens a door to your computer — and malefactors try to push through that door to access your computer. Windows 11 has a *firewall* that monitors and restricts traffic in and out. Don't be afraid of connecting to the Internet, but be aware of the risks and be careful to connect to a network that seems trustworthy. In Chapter 1, you create a local user account. You need a Microsoft Account to take full advantage of Windows 11 features such as the Microsoft Store for apps (see Chapter 10), OneDrive for online storage (see Chapter 16), and synchronized settings between computers. In this chapter, you create a Microsoft Account and choose a secure method for logging in to your account. You also discover how to switch to Airplane mode, create a local account, and switch from a local account to a Microsoft Account. To control access to your computer, you find out how to use a password to unlock your computer.

TIP

If other people use your computer, you may want to create more than one local account. When each person who uses your computer has a separate account, you can keep data, apps, and settings tidy and private. We recommend having only one administrator account for each computer.

Even if you're the only one using your computer, you may want more than one account. For example, if you create a local account, you can experiment with the new account — changing the look and function of Windows 11 — without affecting your first account.

TIP

Many of the steps in this chapter involve entering text, such as your name. If you don't have a physical keyboard, use the virtual keyboard, which is covered in Chapter 1.

Connect to the Internet

1. On the Windows desktop, look to the Network icon to see whether your computer is connected to the Internet. This icon is located to the left of the time and date readings in the lower-right corner of the screen on the taskbar, as shown in **Figure 4-1.**

TIP

Select or choose options by moving the mouse pointer and left-clicking or by tapping a touchscreen with your finger.

2. Select the Network icon. A network panel appears on the right side of your screen, as shown in **Figure 4-2,** and shows what network you are connected to. Select the Wi-Fi button to see available networks, as shown in **Figure 4-3.**

TIP

If you see *Not connected* and *No connections available,* you may be out of luck. Check your computer documentation to see whether your PC has wireless capability and whether you need to turn on a mechanical switch.

TIP

If your computer is near a router (DSL or cable) and you don't have wireless capability, you may be able to connect your PC and the router using an Ethernet cable, which is thicker than a phone line, with wider connections.

3. Select a connection. Note that not all displayed connections are accessible or desirable.

TIP

If you're not sure that you can trust a connection, you might want to forego a connection — better safe than sorry. (Unsafe wireless connections can be used to eavesdrop on your activities, though that scenario is rare.) However, if an available connection sports the name of the establishment you're in or near, such as a restaurant or a coffee shop, the connection may be safe. The same is true of connections at libraries, airports, and many other public spaces.

4. After you select a connection, the selected icon expands and displays the Connect Automatically check box, as shown in **Figure 4-4.** If you trust the connection and might want to use it again, select the check box. Otherwise, deselect the check box. To continue, select the Connect button.

5. You may be prompted to enter a network security key (a series of characters), which limits access to those who know the key. See **Figure 4-5.** The key protects that network and its users. If you're using a hotel's connection, you can obtain the key from the front desk. If you don't know the key, select Cancel. Otherwise, enter the key (dots appear as you type) and select Next.

TIP

If the connection is public and open, you won't be prompted for a key. Open connections are common in libraries, coffee shops, and other places many people come and go.

6. If you entered the correct key or none was required, you may see this message: *Do you want to turn on sharing between PCs and connect to devices on this network?* The term *sharing* refers to allowing computers access to your files or to a device, such as a printer. You should share at home but not in a public location. Choose one of the following:

- **No:** This option prevents you from connecting directly with other computers and protects your computer from networks in public places. You'll still have Internet access.

- **Yes:** This option enables you to share documents and devices between your computers on a home or office network.

TIP

If you're not sure about which option to choose, go with No.

7. When a Wi-Fi connection is established, the word *Connected* appears next to the network name in the Networks panel, as shown in **Figure 4-6.** The connection name and signal strength appear as well. The connection strength (but not the name) appears near the time and date in the notification area in the lower-right corner of the screen (see **Figure 4-1**).

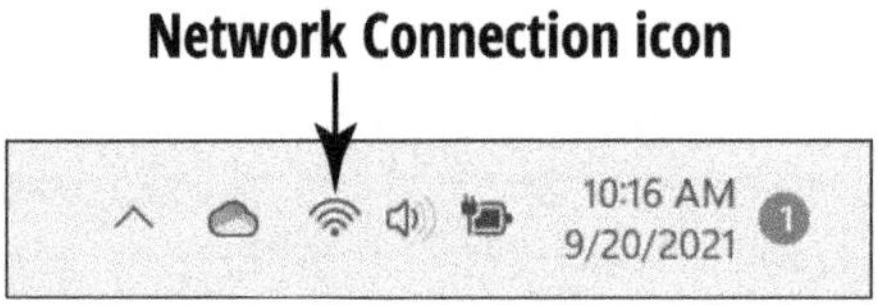

Network Connection icon
↓

| ∧ ☁ 🛜 ◁» 🔋 | 10:16 AM
9/20/2021 | ① |

FIGURE 4-1

Network Connection

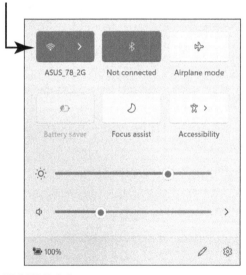

FIGURE 4-2

FIGURE 4-3

FIGURE 4-4

TIP

If you selected the Connect Automatically check box (in Step 4), the connection will be used anytime it is available. If you move your computer to another location out of range of this network (usually a few hundred yards), you will have to repeat these steps to connect to another network.

FIGURE 4-5

FIGURE 4-6

Disconnect (or Switch to Airplane Mode)

1. When you shut down your computer or move your computer far enough away from the connection, your computer disconnects from the Internet automatically. Suppose you want to disconnect on your own? To disconnect your computer from the Internet, select the network icon on the taskbar to display the network panel again (refer to **Figure 4-2**).

2. To turn off Wi-Fi, just select the Wi-Fi button (the button changes from blue to white). To reconnect, select the Wi-Fi button again (the button changes back to blue). (If you disconnect your computer, reconnect before continuing to Step 3.)

3. For safety's sake, airlines don't want passengers to send or receive wireless data signals while the airplane is in flight; these signals can interfere with the airplane's communications systems. This is why the captain gives the order to "turn off all electronic devices." Rather than turn off your computer, however, you can switch to Airplane mode. To do so, open the network panel and select the Airplane Mode button. This button is located at the top right of the network panel.

4. In Airplane mode, an airplane icon appears to the left of the time and date readings in the lower-right corner of the screen, as shown in **Figure 4-7.** Your computer doesn't send wireless data signals in Airplane mode. To switch out of Airplane mode and reconnect to the Internet, select the airplane icon to display the network panel. Then select the Airplane Mode button (refer to **Figure 4-2**).

Airplane mode

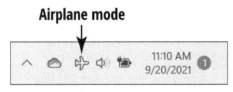

FIGURE 4-7

 TIP Switching to Airplane mode is much more convenient than turning off your computer's network connection and turning it back on again. It takes but a second or two.

Create a New Microsoft Account

1. On the Start screen, select your picture on the left side of the screen. In the pop-up menu that appears, choose Change Account Settings, as shown in **Figure 4-8.** The Settings app opens to the Accounts screen, as shown in **Figure 4-9.** If your account information appears

with the words *Local Account* under your name, you are signed in with a local account. If you don't see Local Account, you already have a Microsoft Account and don't need to follow these steps.

2. Select the Sign In with a Microsoft Account Instead link. The Sign In screen appears. To create a new account, select the Create One! link.

3. On the Create Account screen, create the account with an email address or a phone number. Messages about your Microsoft Account will be sent to the email address or phone number you enter.

   - **Email address:** Type the email address you want to use for this account in the Email Address box and select Next.

   - **Phone Number:** Select the Use a Phone Number Instead link. Then enter your phone number and select Next.

4. In the Create a Password box, type a password. Dots appear instead of what you type. Use at least eight characters, including at least two uppercase letters, two lowercase letters, two numbers, or two symbols. No spaces are allowed.

TIP

Your password should be easy enough for you to remember and type, but not easy for someone else to guess. Don't use the names of any family members (including pets). A good password can be difficult to create, so try this trick: Think of a memorable phrase or lyric. Next, use the first letter from each word, capitalizing some of those letters. Then substitute some letters with numbers (for example, 3 for the letter *E* or the number 0 for the letter *O*). Don't put your password on a sticky note near your screen.

5. Enter your first and last name; then select Next.

6. In the What's Your Birth Date? screen, enter your country or region if it is not already selected.

7. Select the month, day, and year of your birth. (You can lie, but be sure to remember your response.)

8. Review your entries and select the Next button.

9. On the Verify screen, enter the code Microsoft sent to the email address or phone number you entered in Step 3.

10. Select the Next button. The Sign In to This Device Using Your Microsoft Account screen appears. Enter the user name and password to your local account, if you had a password. Then click Next. Congratulations! You just created a Microsoft Account.

Depending on your computer, you may have other sign-in options called Windows Hello, such as facial recognition, fingerprint recognition, and PIN sign-in. These options may or may not be available depending on your computer's hardware. However, you can check them out by accessing Accounts⇨ Sign-in options.

FIGURE 4-8

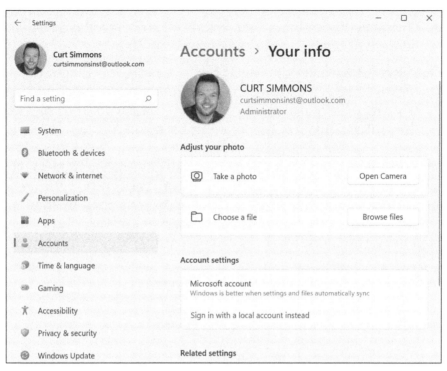

FIGURE 4-9

Create a Local Account

1. "Local account" is Microsoft's name for your ability to start your computer without signing in to Microsoft. Before you create or sign in with a local account, save your work. Then select the Start button, select your picture on the Start menu, and in the pop-up menu that appears (refer to **Figure 4-8**), choose *Sign in with a local account instead*.

TIP

You can't download and install apps from the Microsoft Store without signing in with a Microsoft Account. Nor can you upload files to OneDrive. So why have a local account? If you're concerned about privacy and security, if you don't want Microsoft to peer over your shoulder and track your every move when you use your computer, consider creating a local account. You can always switch to your Microsoft Account when you want to install apps from the Microsoft Store, use OneDrive, or use Skype (see "Switch from a Local to an Existing Microsoft Account," later in this chapter).

TIP

To create a local account for someone else to use your computer, see "Create a Local Account for Someone in Your Household," later in this chapter.

2. Choose the Sign In with a Local Account Instead link. You see the Switch to a Local Account screen, shown in **Figure 4-10.** Select the Skip This Step button and select Next on the following screen.

3. Enter the password of your Microsoft Account on the Making Sure It's You screen, shown in **Figure 4-11,** and select OK.

4. In the next screen, shown in **Figure 4-12,** enter a user name, a password (twice), and a word or two to jar your memory if you forget your password; then select Next.

TIP

Passwords are case sensitive. In other words, your computer recognizes uppercase letters and lowercase letters when you create and enter a password. Remember which letters in your password are upper- and lowercase.

TIP

If you want to sign in to your local account without having to enter a password, leave the Password, Reenter Password, and Password Hint text boxes blank. Later in this chapter, the sections "Create a Password for a Local Account" and "Change or Remove a Local Account Password" explain the ins and outs of local account passwords.

5. Select the Sign Out and Finish button. Your Microsoft Account closes, Windows restarts, and you see the Windows 11 screen.

6. Enter the password you created in Step 3. When you start your computer, Windows 11 assumes that you want to sign in with the account you were using when you last shut down your computer. If you shut down when signed in with your local account, you are asked to provide the password for your local account the next time you start your computer.

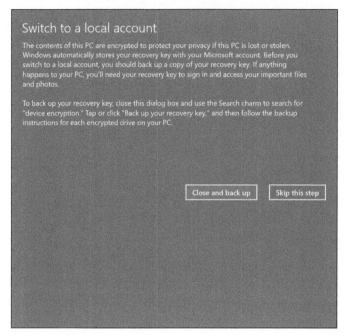

FIGURE 4-10

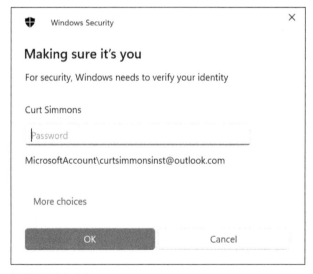

FIGURE 4-11

Enter your local account info

Enter the user name and password for your new or existing local account. This is the account you'll use to sign in to Windows from now on.

If you want to continue to sign in to Windows with Windows Hello, you'll need to set up a password.

User name   curts   ✕

New password

[Next]   [Cancel]

FIGURE 4-12

Switch from a Local to an Existing Microsoft Account

1. Follow these steps to sign in with your Microsoft Account when you are currently signed in with your local account. On the Start screen, select your name in the bottom-left corner of the screen and then choose Change Account Settings from the pop-up menu that appears (refer to **Figure 4-8**). The Accounts screen opens. Your account information appears with *Local Account* under your name, shown in **Figure 4-13.**

TIP If you don't have a Microsoft Account already, see "Create a New Microsoft Account," earlier in this chapter.

TIP If you don't see *Local Account* under your user account name, you're already signed in with your Microsoft Account.

2. Select the Sign In with a Microsoft Account Instead link. The screen shown in **Figure 4-14** appears. Type the email address or phone number of your Microsoft Account and select the Next button. Note that you must be connected to the Internet to sign in with a Microsoft account.

3. Enter the password of your Microsoft Account and select the Sign In button.

4. In the Current Windows Password text box, enter the password of your local account and select the Next button. The Accounts screen opens, where you see the user name and email address of your Microsoft Account.

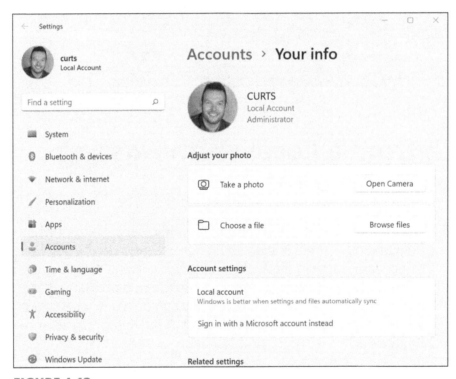

FIGURE 4-13

FIGURE 4-14

Create a Local Account for Someone in Your Household

1. Consider creating a local account for each person who uses your computer. A local account is more than just a way to sign in. Each local account preserves the user's settings and Internet browsing history. When you set up a local account, Windows 11 creates separate folders called Documents, Pictures, Videos, and Music for the account holder. Only the account holder can open and edit files in these special folders. To create a local account, select your name on the Start screen and choose Change Account Settings on the menu that appears (refer to **Figure 4-8**). The Settings app opens to the Accounts screen.

TIP

To create a local account, you must be your computer's administrator. To see whether you're the administrator, look for the word *Administrator* below your name on the Your Info page of the Settings screen.

2. On the left, select Accounts. On the right, select Family & Other Users. On the next screen that appears, select Add account next to Add Other User, shown in **Figure 4-15.** The How Will This Person Sign In screen appears.

3. At the bottom of the How Will This Person Sign In screen, select the link called I Don't Have This Person's Sign-In Information.

TIP

If you know you want a Microsoft Account instead of a local account, jump to "Create a New Microsoft Account" or "Switch from a Local to an Existing Microsoft Account," earlier in this chapter.

4. On the next screen, select Add a User without a Microsoft Account.

5. On the next screen (it's called Create a User for This PC), enter the user name in the first box, as shown in **Figure 4-16.** Use the person's first name, last name, first and last name, initials, or nickname — something easy to remember and type.

TIP

You are not required to use a password with a local account, which makes signing in easy. However, without a password, anyone can use the computer and access information that you might want to protect.

6. In the Password box, enter a password (or skip to Step 9 if you don't care to create a password for your account). A dot will appear for each character you type.

TIP

For suggestions on creating a good password, see the tip in Step 4 in the section, "Create a New Microsoft Account," earlier in this chapter.

7. In the Reenter Password box, type the same password exactly.

8. In the three drop-down menus, choose a security question and type the answer in the text box. If you forgot your password, you will be able to recover it by answering a security question correctly.

9. After completing all the available boxes, choose the Next button. In the Accounts screen, the new user name appears under Other Users, as shown in **Figure 4-17.**

10. Return to the Start screen and select your user name. Notice that the new user name appears on the drop-down menu. You can switch between accounts by selecting a user name on the drop-down menu. To sign out of an account, choose Sign Out.

11. In the drop-down menu, select the new user name to switch to that account.

12. A screen appears with the new user name. If you used a password on the new user account, type that password in the box and select the onscreen right arrow or press Enter. If you didn't use a password, select the Sign In button.

FIGURE 4-15

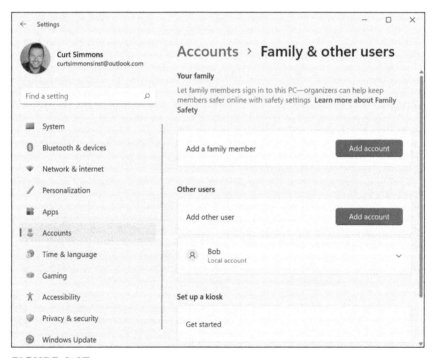

FIGURE 4-16

FIGURE 4-17

TIP

The first time you sign in as a new user, you have to wait a moment while apps are installed for the new user. Soon the generic Start screen appears. (Any settings you changed in your account do not transfer to other accounts.)

TIP

When you start a computer for which you've created more than one user account, the users' names appear in the lower-left corner of the Windows 11 screen. To tell Windows which user to sign in with, select a user name on the Windows 11 screen before signing in.

Create a Password for a Local Account

1. If necessary, sign in to the local account that needs a password. On the Start screen, select your name. From the drop-down menu that appears, choose Change Account Settings (refer to **Figure 4-8**). The Settings app opens to the Accounts screen.

TIP

If you already have a password, see the "Change or Remove a Local Account Password," later in this chapter.

2. Select Sign-in Options.

3. Select Password.

4. If you don't have a password but want one, select the Add button under Password.

TIP

The buttons available under Sign-in Options depend on your current setup. You may see buttons that enable you to create, change, or remove a particular setting.

5. On the Create a Password screen, enter a password in the New Password box.

TIP

For a local account, the password can be any length. See Step 4 in the "Create a New Microsoft Account" section for suggestions about creating a good password.

6. In the Reenter Password box, enter the password again.

7. Enter a hint to remind yourself — and no one else — about your password. And whatever you do, don't include the password in the hint.

8. Select Finish. If any error messages appear, correct the entries and select Next again.

Change or Remove a Local Account Password

1. If you haven't done so already, sign in to the local account with the password you want to change or remove. On the Start screen, select your name and then choose Change Account Settings. On the left side of the Accounts screen, select Sign-in Options.

2. In the middle of the screen, select Password.

3. Select the Change button.

TIP

If you don't have a password but want one, see the preceding "Create a Password for a Local Account" section.

4. On the Change Your Password screen, enter your current password and then select the Next button.

5. On the next Change Your Password screen enter the new password. (If you don't care to have a password, select Next and skip to Step 8.)

TIP

To remove your current password and use no password, leave all boxes blank. However, especially if you have a laptop that you carry with you, going without a password isn't recommended. Without a password to safeguard it, anyone can get into your laptop.

6. In the Reenter Password box, enter the password again.

7. Enter a hint to remind yourself — and no one else — about your password. Then select Next. If any error messages appear, correct the entries and select Next again.

8. The final screen indicates that you must use your new password the next time you sign in. (This message appears even if you left the password blank, in which case you won't need any password.) Select Finish.

Delete a Local Account

1. Before you delete a local account, make sure that the user of that account is signed out. Moreover, you must be signed in to a Microsoft Account to delete a local account.

2. Select your name on the Start screen and choose Change Account Settings. You land on the Your Info page of the Settings screen. Does it say "administrator" under your name? We hope so, because you must be your computer's administrator to delete a local account. On the Accounts screen, select Family & Other Users.

3. Select the account you want to delete and then select the Remove button, as shown in **Figure 4-18.**

4. In the Delete Account and Data window, select the Delete Account and Data button.

TIP

Take heed of the onscreen warning. Deleting a user account removes the user's data, including all documents, music, and other files. If you're not sure which option is best, choose Cancel.

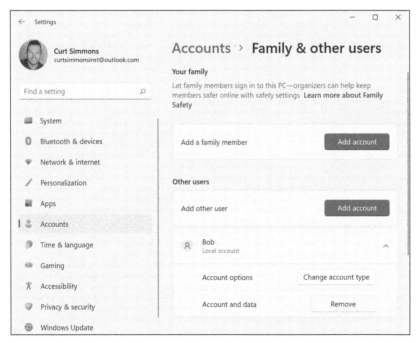

FIGURE 4-18

Chapter **5**

Getting Comfortable with the Desktop

The *desktop* is Grand Central Station as far as Windows 11 is concerned. Sure, you can open applications from the Start screen (Chapter 2 explains how), but with a few simple modifications, you can open applications from the desktop as well. When you're running more than one application, you can go to the desktop and quickly switch from one application to another. In fact, in Windows 11, you can create a second "virtual" desktop for one set of open applications (applications that pertain to leisure, let's say), and when the boss isn't looking, you can switch to the second desktop to play games or chat with friends.

One key feature of the desktop is the *taskbar,* a strip along the bottom of the screen that shows icons for desktop programs. The taskbar can be used to run and switch between desktop programs.

In this chapter, you get acquainted with the desktop, the taskbar, and apps. You change the date, time, and time zone, as needed. You resize and reposition windowed apps, and discover how to "add a desktop" for a second set of open applications. You select a background for the desktop and make some desktop apps more convenient to use by pinning them to the taskbar. Finally, you work with the Task Manager, which lets you end any app that is misbehaving.

The desktop originated when using the mouse was the most common method for selecting objects (touchscreens were nonexistent). Therefore, on the desktop, a few tasks are easier to do with the mouse than with touch or a keyboard.

TIP

This chapter is an introduction to the desktop. See Part 4 to dive a little deeper into desktop functions, such as organizing documents.

TIP

Check Out the Desktop

1. Go to the desktop (if you aren't already there) by using one of these techniques:

- Press ⊞+D.

- Right-click the Windows button and choose Desktop on the pop-up menu that appears.

2. Your desktop has a picture in the background. Examine your desktop for *icons* — small pictures that represent either programs, which perform functions, or documents, such as letters and photos. You select an icon to run a program or open a document. The Windows 11 desktop displays an icon for the Recycle Bin, where deleted documents go. The Recycle Bin may be the only icon on your desktop, or you may see others. See Chapter 15 for information on using the Recycle Bin.

3. The area at the bottom of the screen is the *taskbar,* shown in **Figure 5-1.** From left to right, the taskbar offers these features:

- **Start button:** Selecting this button opens the Start screen. (Click or tap the button a second time to close the Start screen.)

TIP

- **Search button:** You can select Search to open the Search screen and enter a search term here to look for Windows settings, applications, files on your computer, and information on the Internet.

- **Virtual Desktop:** Select this button to create an additional virtual desktops (see "Open a Second Desktop," later in this chapter, for more information).

- **Icons:** Some icons appear on the taskbar automatically. Very likely, your taskbar has icons for File Explorer and Edge. You can select these icons to open File Explorer and Edge. When you open an application, Windows 11 places its icon on the taskbar.

- **System tray:** The *system tray* displays icons for programs that run automatically when your computer starts. The date and time are to the right of the icon tray. On the right side of the icon tray is the Notification icon.

You can use the taskbar to switch among programs by selecting the icon for the program you want to use.

4. Select an icon on the taskbar to open the associated program. (Refer to **Figure 5-1** if you aren't sure where the taskbar is located.)

You select items on the desktop or in the taskbar by clicking with the mouse or tapping the touchscreen. To some extent, you can use the Tab and arrow keys, but that's an awkward method.

TIP

If you have a touchscreen, note the taskbar icon for the virtual keyboard. See Chapter 1 for information about using the keyboard.

TIP

5. Right-click over an icon or tap and hold until a small box appears, and then release. A *context menu* appears with options specific to the icon you selected, as shown in **Figure 5-2.** Select anywhere else on the desktop to dismiss this menu. Repeat this on a few different areas of the screen to see how the context menu changes.

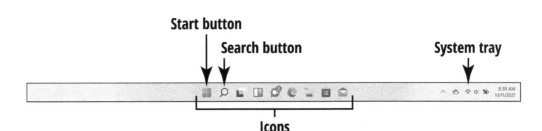

FIGURE 5-1

Apps and Features

Mobility Center

Power Options

Event Viewer

Device Manager

Network Connections

Disk Management

Computer Management

Windows Terminal

Windows Terminal (Admin)

Task Manager

Settings

File Explorer

Search

Run

Shut down or sign out   >

Desktop

FIGURE 5-2

Change the Date or Time

1. Select the date and time displayed on the taskbar. A calendar pops up, as shown in **Figure 5-3.**

2. If the date or time is incorrect for your location, select the Search button on the taskbar and then search for Date and Time settings. Select the option in the search results. You can also right-click the date and time in the system tray and select Adjust Date & Time. Either way, the Settings screen opens. Select Time & Language. You see the Date & Time window, shown in **Figure 5-4.**

3. Windows 11 determines the correct time and date from the Internet, and your computer should show the right time and date, but if it doesn't, turn off the Set Time Automatically option and select the Change button. You see the Change Date and Time screen (see **Figure 5-5**). Select the correct date and time on this screen. Change the time by using the little triangles that point up (later) or down (earlier) or by entering the specific hours and minutes. Select Change to keep your change or Cancel to ignore your change.

4. Back in the Date and Time window, select your Time Zone from the drop-down list, if necessary. Turn the Adjust for Daylight Saving Time option on or off as appropriate.

5. Note the toggle button for Adjust for Daylight Saving Time Automatically. If you live in an area where daylight saving time is in effect, keep this button turned on so that your computer will automatically adjust for daylight saving time.

FIGURE 5-3

Turn off automatic time   **Turn daylight saving time on or off**

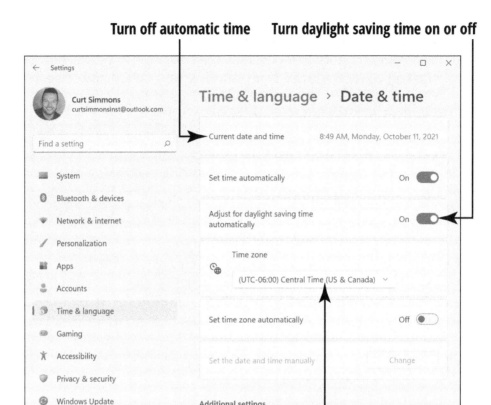

Adjust your time zone

FIGURE 5-4

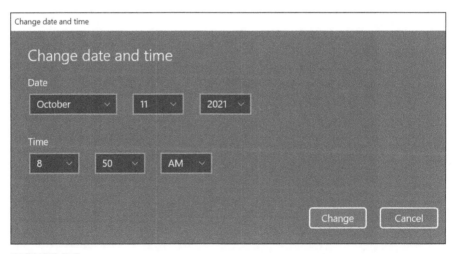

FIGURE 5-5

Explore the Parts of a Window

1. On the taskbar, select the File Explorer icon (it looks like a folder). File Explorer opens, as shown in **Figure 5-6.**

TIP

File Explorer enables you to view your computer storage, such as hard drives, and folders, which are used to organize your documents. See Chapter 15 for information on using File Explorer.

2. Explore the example window in **Figure 5-6,** starting at the top left:

- **Title bar:** The *title bar,* which is the top line of the window, lists the name of the file or folder that is currently open.

    TIP

    The title of the window in **Figure 5-6** is This PC, which is the location File Explorer is focused on when you open File Explorer.

- **Toolbar:** The toolbar gives you access to common tools, such as Cut, Copy, Paste, Sort, View, and others.

- **Minimize:** The *Minimize button,* located in the upper-right corner of the window, shrinks or hides the window's contents. The program that the window contains is still running and open, but the window is out of sight. You'll still see the program's icon on the taskbar. Select the Minimize button when you want to ignore a particular window but aren't actually done with it. To restore the window, select its icon on the taskbar.

- **Maximize/Restore:** The *Maximize button* (the button with a single square in the upper-right corner of the window) fills the screen with the contents of the window. Select the Maximize button to hide the desktop and other open windows, to concentrate on one window, and to see as much of the window's contents as you can. The *Restore button* (the button with two squares in the upper-right corner) is the name of the button that appears after you select the Maximize button; it replaces the Maximize button. Select the Restore button to return the window to its previous size, which is between maximized and minimized. (Press ⊞+up-arrow key to maximize, and ⊞+down-arrow key to restore or minimize.)

- **Close:** The *Close button* is the button with the X in the upper-right corner of the window. Select the Close button when you are done with the window. Close is also called Quit and Exit. (Press Alt+F4 to close the current window or the desktop itself. This keyboard shortcut works for Windows 11 apps, as well.)

- **Window Contents:** The bulk of the window contains the program or document you're using. File Explorer displays locations on the left and objects in that location on the right.

- **Status bar:** Along the bottom edge of the window, some programs display information about the window or its contents in a single-line *status bar*. File Explorer lists how many files are in the currently open folder and how many files (if any) have been selected.

  Scan the edges of windows. Often, important information and functions are pushed to these edges around the main content area.

3. Select the Close button (the X) to close File Explorer.

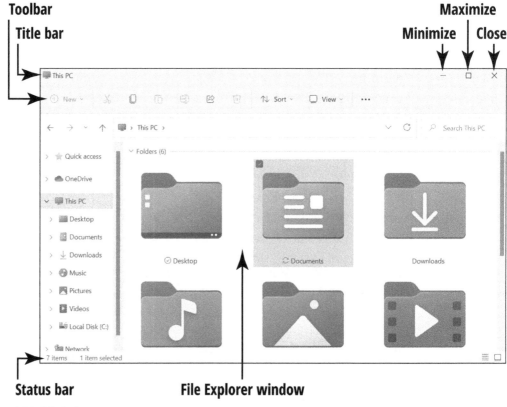

FIGURE 5-6

TIP

See Chapter 2 for information on finding other desktop programs, such as the Calculator.

Resize a Window

1. To resize a window, open File Explorer by selecting the folder icon in the taskbar. (Refer to **Figure 5-6**.)

 2. If the window is maximized (fills the screen), select the Restore button to make the window smaller.

3. Use one of these methods to resize the window:

   - **Mouse:** Move the mouse pointer to the right edge of the window until the pointer changes to a double-headed arrow, called the *resize pointer*. Click and drag the edge of the window, using the resize pointer. (To drag, click and hold down the mouse button while you move the mouse.)

   - **Touchscreen:** Drag the right edge of the window.

     Drag left to shrink the window and right to expand it.

4. Resize the window's width and height at the same time by dragging a corner of the window. If you want a challenge, try resizing the top-right corner without accidentally selecting the Close button.

5. Resize the window's width *or* height by dragging any of the four sides.

   You may want to resize a window to show only what you need to see, nothing more. Practice resizing from any side or corner.

6. Leave the window open as you go on to the next task.

Arrange Some Windows

1. On the desktop, select and open the Recycle Bin by double-clicking or double-tapping its icon. The Recycle Bin contains deleted files and folders. It appears in another File Explorer window. See **Figure 5-7.**

   Double-click by clicking the left mouse button twice, without a pause. Double-tap by tapping twice in quick succession.

2. If File Explorer isn't still open from the preceding section, open it by selecting the folder icon on the taskbar. You now see two overlapping windows on the desktop (refer to **Figure 5-7**), one titled *This PC* and the other titled *Recycle Bin*.

The window in front of the others is called the *active* window. All other windows are *inactive.* Selecting anywhere in an inactive window makes it active and moves it to the front of the others.

3. Drag the Recycle Bin title bar (avoiding the buttons on the left and right ends) to move that window a little.

4. Drag the This PC title bar (again, avoiding the buttons on both ends). The This PC window moves in front of the Recycle Bin as you move it. Move both windows so that you can see a little of each (refer to **Figure 5-7**).

5. Practice moving both windows. Arranging windows helps you see and do more than one thing at a time. Use the techniques from the preceding section, "Resize a Window," to see as much as you can of both windows at one time.

If you can't see the title bar of the window you want to move, move the other window to uncover the hidden title bar.

6. Leave both windows open for the following task.

Move a Window by dragging its title bar

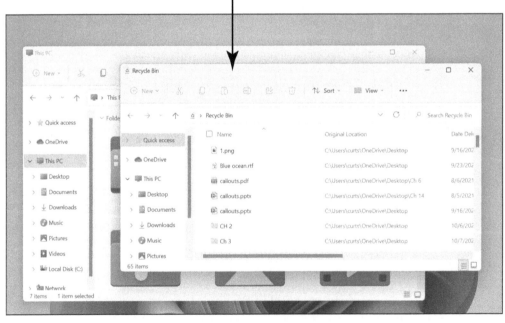

FIGURE 5-7

Use Snap Layouts

1. With the File Explorer window and Recycle Bin open, also open the Edge browser and the Windows Store (both are available as icons on the taskbar).

2. Now you have four different windows open on your desktop. That's a lot of windows to manage. What if you're trying to work with all of them at the same time? You would have to toggle back and forth between those windows, which would be time consuming and confusing. Windows 11 provides an easy and fun way to work with several windows at the same time on the desktop, called Snap Layouts. Point your mouse to the Maximize button on one of the windows (it doesn't matter which one) and you see a pop-up menu appear, as shown in **Figure 5-8.**

3. Each snap layout group represents an area of the desktop where you can snap the selected window to. For example, choose the left block in the first snap group and then choose the right block in the first snap group for a second window. This snaps both windows in place so that you can easily work with them at the same time, as shown in **Figure 5-9.**

4. What if you have four open windows? In that case, just choose the four-window snap layout and snap each open window to the desired location. For example in **Figure 5-10,** File Explorer, Recycle Bin, Edge, and the Windows Store are snapped into a four-grid layout.

5. In the snap view, you can work with each window without minimizing the others. To move a window out of a snap view, just drag its title bar or select the Maximize button. Then you can move the window around, resize, or close it as you normally would.

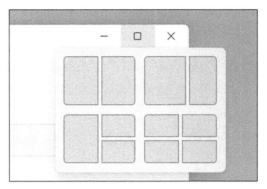

FIGURE 5-8

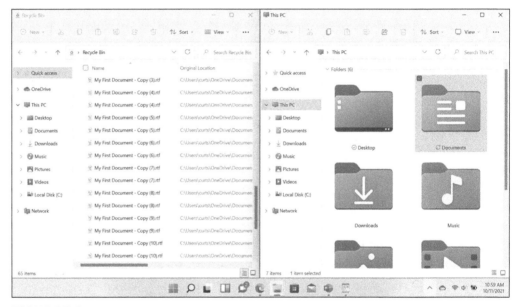

FIGURE 5-9

TIP

You can maximize a single window by dragging it to the top edge of the screen. Dragging a window to the top is the equivalent of selecting the Maximize button (see the section "Explore the Parts of a Window," earlier in this chapter).

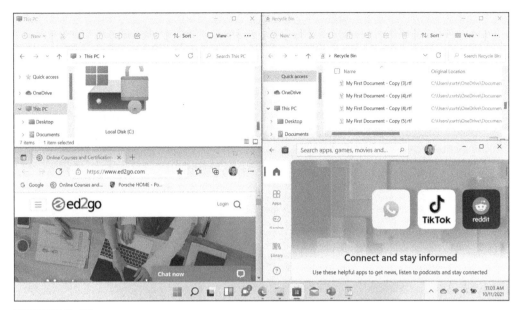

FIGURE 5-10

Open a Second Desktop

1. If File Explorer and Recycle Bin aren't open on your desktop (because you didn't follow the previous exercise in this chapter), open those programs now.

2. Select the Virtual Desktop button on the taskbar. After you select the Virtual Desktop button, a thumbnail version of an open desktop appears and you see an option to select New Desktop, as shown in **Figure 5-11.**

3. Select New Desktop. You see a brand-new, pristine desktop, labeled Desktop 2, as shown in **Figure 5-12.** Congratulations; you are now in Desktop 2. Create a second (or third or fourth) desktop when you want to keep the first desktop from getting crowded with too many applications or to better organize your work. For example, if you're using applications for your work and using other applications for leisure purposes, put the work applications on one desktop and the leisure applications on another. This will help you find the application you want to work with.

4. While you are on Desktop 2, open the Photos app. Having just one application on the desktop makes working with that application a little easier.

5. Move the pointer or your finger onto the Desktop 2 tile, and when the Close button appears, click or tap the Close button to close Desktop 2. The Photos app that you opened on Desktop 2 now appears on Desktop 1, the only open desktop. When you close a desktop, all its open applications move to the desktop that is still open.

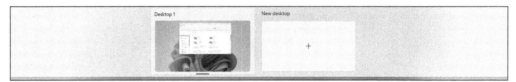

FIGURE 5-11

FIGURE 5-12

Choose a Desktop Background

1. Select the Search icon on the taskbar and search for Settings. Choose the Settings app in the search results.

2. Select Personalization. Then, in the Personalization window, select Background. The Background screen, shown in **Figure 5-13,** is the place to go to choose a background for the Windows desktop.

TIP

You can use the Personalization window to customize many aspects of the desktop. The more time you spend on the desktop, the more worthwhile this personalization may be.

3. In the Personalization window, select Picture on the Background menu (see **Figure 5-13**).

4. Select any photo to make that photo the desktop background. The background changes immediately. To see the entire desktop, minimize the Settings window. Restore the Settings window by selecting its icon on the taskbar or by repeating the preceding steps.

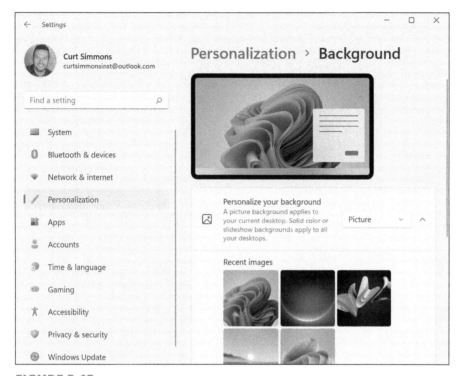

FIGURE 5-13

To use a photo of your own as the desktop background, select the Browse Photos button and choose a photo in the Open dialog box.

Note on the Personalization window, you choose Themes and apply a different theme to your Windows 11 PC. Doing so will affect the desktop photo, colors, and possibly even sounds. You can also download free themes from the Microsoft Store to make Windows 11 even more personalized. See Chapter 10 to find out more about the Microsoft Store.

Pin Icons to the Taskbar

1. On the Start screen, select the All Apps button and locate the Calculator icon, as shown in **Figure 5-14.**

2. Use one of these methods to display the shortcut menu on the Calculator icon:

   • **Mouse:** Right-click the Calculator icon.

   • **Touchscreen:** Swipe the Calculator icon down or up slightly.

3. Select More on the shortcut menu. A submenu with more options appears (see **Figure 5-15**).

4. Select Pin to Taskbar on the submenu. Doing so places the icon for Calculator on the desktop taskbar for easy access. The Calculator icon appears on the taskbar.

TIP

Press +D to go directly to the desktop at any time.

FIGURE 5-14

FIGURE 5-15

5. Repeat Steps 1 and 3 and then select Unpin from Taskbar.

A fast way to remove an icon from the taskbar is to right-click it and choose Unpin from Taskbar on the shortcut menu.

TIP

6. Switch to the desktop. The Calculator icon is gone.

Pinned icons have a killer feature: *jumplists,* which are short menus of tasks and documents specific to the pinned app. To see the jumplist of a pinned app, right-click its icon on the taskbar or tap and hold on the icon until a small box appears. Try that with the File Explorer icon on the taskbar. Not all desktop apps have jumplists.

TIP

Stop Apps with the Task Manager

1. Select the Search icon on the taskbar and search for Task Manager. Select Task Manager from the search results.

2. The Task Manager lists any running apps — both desktop apps and Windows 11 apps. Select an app, also referred to as a *task.* Note that the End Task button is now available, as shown in **Figure 5-16.** You don't have to end this task, but you could. Any of the tasks in the Task Manager window can be ended without consequences.

TIP

Be careful about ending an app used to create something (for example, a word-processing app) because you could lose data you haven't saved before ending the task. Use the Task Manager to end tasks that you can't end otherwise, such as a frozen or locked app or one that seems to slow down everything.

3. Close the Task Manager.

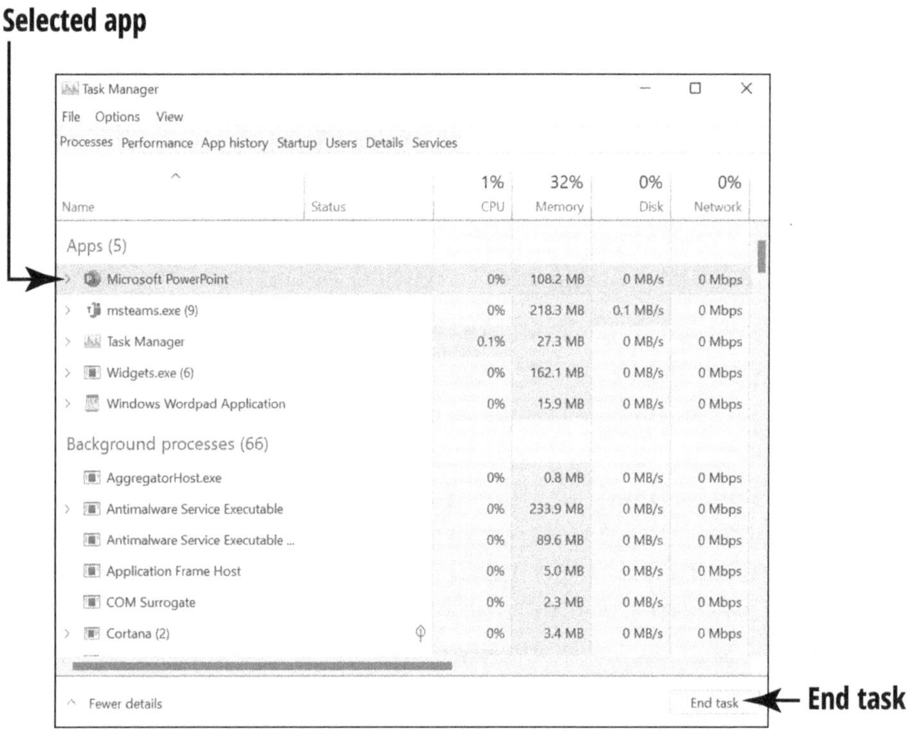

Selected app

Task Manager			1%	32%	0%	0%
File Options View						
Processes Performance App history Startup Users Details Services						
Name	Status		CPU	Memory	Disk	Network
Apps (5)						
Microsoft PowerPoint			0%	108.2 MB	0 MB/s	0 Mbps
msteams.exe (9)			0%	218.3 MB	0.1 MB/s	0 Mbps
Task Manager			0.1%	27.3 MB	0 MB/s	0 Mbps
Widgets.exe (6)			0%	162.1 MB	0 MB/s	0 Mbps
Windows Wordpad Application			0%	15.9 MB	0 MB/s	0 Mbps
Background processes (66)						
AggregatorHost.exe			0%	0.8 MB	0 MB/s	0 Mbps
Antimalware Service Executable			0%	233.9 MB	0 MB/s	0 Mbps
Antimalware Service Executable ...			0%	89.6 MB	0 MB/s	0 Mbps
Application Frame Host			0%	5.0 MB	0 MB/s	0 Mbps
COM Surrogate			0%	2.3 MB	0 MB/s	0 Mbps
Cortana (2)			0%	3.4 MB	0 MB/s	0 Mbps
^ Fewer details					End task	**End task**

FIGURE 5-16

2

Windows 11 and the Web

IN THIS PART . . .

Explore the Internet.

Send and receive email.

Connect with Microsoft Teams.

Test-drive some valuable apps.

Chapter **6**

Finding What You Need on the Web

The World Wide Web — or, simply, the web — provides quick access to information and entertainment worldwide. One part library, one part marketplace, and one part soapbox, the web makes everything equidistant. From down the block to halfway around the world — even out into space — everything is a few clicks or taps away. News, shopping, and the electronic equivalent of the town square await you.

You explore the web using a *web browser,* a program designed to make browsing the web easy, enjoyable, and safe. In this chapter, we show how you can use the Edge browser to step beyond your computer into the global village.

You browse *web pages,* which are published by governments, businesses, and individuals — anyone can create web pages. Each web page may consist of a few words or thousands of words and pictures. A web page is part of a larger collection called a *website,* which consists of a group of related web pages published on a topic by an organization or individual. Companies and individuals create websites to organize their related pages.

Pages and sites on the web have some common characteristics:

» **Unique addresses,** which are formally called *URLs* (URL stands for Uniform Resource Locator, in case you're ever on *Jeopardy!*).

» **Connecting links** that move you from page to page when you select them. These *links* (also called *hypertext links* or *hyperlinks*) often appear underlined and blue. Pictures and other graphic images can also be links to other pages. You can tell when a picture or image is a link by moving the pointer over it. The pointer changes from an arrow to a hand when it is over a hyperlink picture or image. Exploring the web using links is easier than typing URLs.

In this chapter, you use Edge to browse the web. To get the most out of browsing, you juggle multiple sites simultaneously and find out how to bookmark your favorite websites. You also find out how to search for almost anything and make use of some nice features in Edge. Finally, this chapter shows how to view web pages without seeing pesky advertisements.

TIP

Edge is by no means the only browser — it's just Microsoft's browser. You can also download and try other browsers if you like, namely Mozilla Firefox (www.mozilla.org) and Google Chrome (www.google.com/chrome/browser). With that said, note that Edge does everything you need, so you don't need to download anything else.

Browse the Web with Edge

1. Open Edge by selecting its icon on the taskbar. Edge opens to the *start page,* as shown in **Figure 6-1.** This page presents websites and news stories that Microsoft thinks are of interest to you based on your browsing history. If you haven't used Edge yet or haven't used it very often, you don't have a browsing history to speak of, and the websites and news stories you see here probably look quite tepid. (Later in this chapter, "Explore Edge Features" explains how to choose what web pages you see when you start Edge.)

TIP

When you start Edge, you may see an error message if you're not connected to the Internet. If so, see Chapter 4 for information on connecting. See "Bookmark and Revisit Your Favorite Websites," later in this chapter, for instructions about choosing a home page.

2. Note the *address bar* at the top of the Edge screen (refer to **Figure 6-1**). In the address bar, type **www.dummies.com.** As you type, search suggestions appear in a drop-down menu. Either select www.dummies.com in the search suggestions or press Enter. The web page for the Dummies Press appears, as shown in **Figure 6-2.**

TIP

If you have trouble finding the address bar, try clicking with the mouse or tapping with your finger near the top of the screen, to the right of the four buttons (Back, Forward, Refresh, and Home). This will make the address bar appear.

3. Select a link on the page with a click or a tap. Where are the links? Just about everywhere. When you move the pointer over a link, it changes from an arrow to a hand — that's how you can tell where the links are.

4. Select the Back button (or press Alt+left arrow) to return to the preceding page. This button (an arrow) is located in the upper-left corner of the screen (refer to **Figure 6-2**). Select the Back button to backtrack and revisit pages.

5. Select the Forward button (or press Alt+right arrow) to move forward to the page you visited in Step 3. The Forward button is next to the Back button (refer to **Figure 6-2**). Edge remembers the pages you visit to make it easy to go forward and back.

TIP

To zoom in and make a web page look bigger, press Ctrl+plus sign. To zoom out, press Ctrl+minus sign.

6. Select the Add This Page to Favorites button on the right side of the address bar (or press Ctrl+D). After you select this button, a Add to Favorites panel appears, as shown in **Figure 6-3.** Select the Add button to add Dummies.com to your Favorites list. Later in this chapter, "Bookmark and Revisit Your Favorite Websites" explains how to go to your favorite websites by selecting sites on the Favorites list.

TIP

You can change the text in the Add to Favorites panel before you select the Add button. However, there's usually no need to change the text unless it's overly long or unclear.

TIP

Browsing the web consists of entering addresses, following links, going forward and back, and revisiting your favorite websites. Relatively simple activities can absorb hours.

7. Keep Edge open if you want to go on to the next topic in this chapter.

Address bar

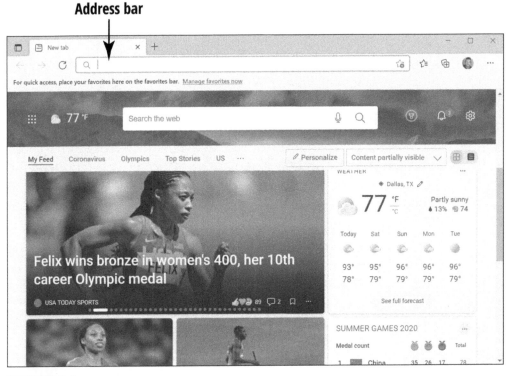

FIGURE 6-1

Back button

Forward button

Add This Page to Favorites

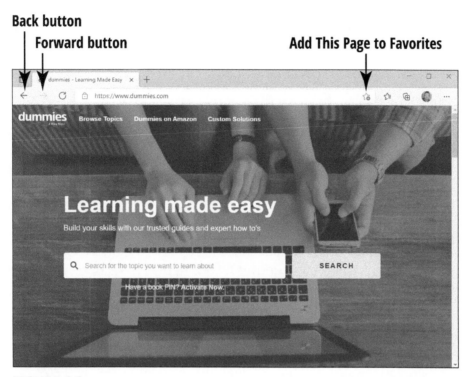

FIGURE 6-2

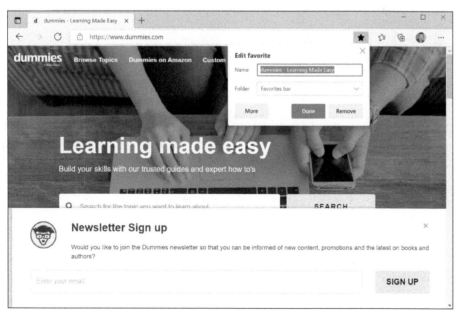

FIGURE 6-3

Open Multiple Pages in Separate Tabs

1. Open Edge if it isn't already open.

2. Go to the Google website at www.google.com. You can get there by typing www.google.com in the address bar and pressing Enter.

3. Select the New Tab button (or press Ctrl+T). This button is located to the right of the rightmost tab, as shown in **Figure 6-4, top.** A new web page appears, as shown in **Figure 6-4, middle.** What's more, a new tab (not coincidentally called New Tab) appears at the top of the screen.

4. On the new tab, enter www.dummies.com in the address bar and press Enter. You open the Dummies website on the second tab, as shown in **Figure 6-4, bottom.** Now two websites are open in Edge. Google.com is open on the first tab; Dummies.com is open on the second. Notice the web page names on the tabs.

TIP

To close a tab, select its Close button (the X) or make sure that you are looking at the page you want to close and press Ctrl+W.

5. Select the first tab, the one with the name Google. You return to the Google website.

Browsing in multiple tabs allows you to keep one page open while visiting another, perhaps to compare information or to follow a different thought.

TIP

The keyboard shortcut for switching between tabs is Ctrl+Tab.

6. Select the Close button (the X) on the Dummies.com tab to close that particular tab. Close tabs when you want to reduce clutter and simplify switching among open tabs.

TIP

The keyboard shortcut for closing the current tab is Ctrl+W. (*W?* Long story.)

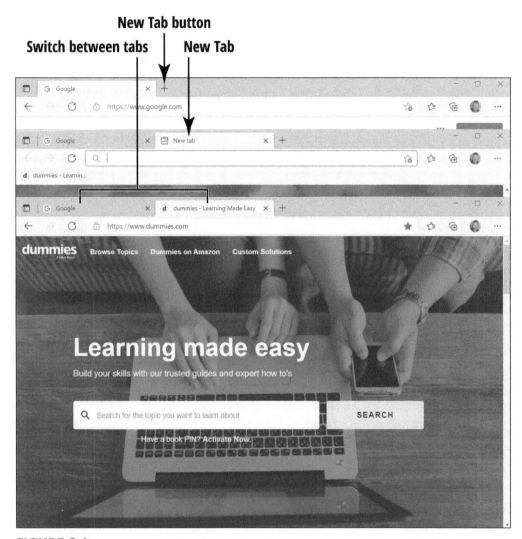

New Tab button

Switch between tabs   **New Tab**

FIGURE 6-4

Search for Anything

1. With Edge open, go to the address bar and type **travel**. A drop-down list of search terms with the word *travel* appears as Edge attempts to guess what you're searching for. (Ignore this list for this exercise, but take advantage of it later.) Select Bing Search on the right side of the search term list to open the Bing website.

TIP

You can search the Internet from the Search icon on the taskbar without opening Edge. Type a search term in the Search box. The Search panel lists items pertaining to the search term you entered, including ideas for web searches. Select a web-related item to search with Edge.

2. A search results page appears, as shown in **Figure 6-5.** The results come from www.bing.com, which is the default search engine for Edge. A *search engine* is simply a website that provides links to web pages that match your search. (That definition, however, ignores the complex process going on behind the scenes.)

3. Scroll down the page of search results. Select any link you want to follow. If you get to the bottom of the page, select the Next button to see more search results.

TIP

Some of the "search results" are advertisements. In Bing searches, the results at the top of the page and the right side of the page are paid for. Be careful of these advertisements, which are designed to sell you something, not to provide information. If you prefer not to see advertisements on web pages, see "Block Ads on Web Pages," later in this chapter.

4. Return to the previous screen by selecting the Back button (or pressing Alt+left arrow).

5. Select the address bar and type **travel new mexico** (no capitals needed). As you type, potential matches for your terms appear in the drop-down list. If you see an item matching the search you want, select it in the list. Otherwise, press Enter.

TIP

Different search engines turn up different results. Other search engines include Google (www.google.com) and Yahoo! (www.yahoo.com). To use one of these search engines when exploring the web, enter its address in the address bar.

6. Note the tabs at the top of the Bing page, below the Search box. The first tab is All, which contains the results you see by default. Additional tabs vary with the search. Select each of the tabs, which may include any of the following:

- **Images** displays pictures matching your terms.
- **Videos** displays clips and snippets related to your search terms.
- **Maps** will help you get there.

- **News** displays search results from recent news, instead of all the results of the broader web.

- **Shopping** directs you where to shop for items that match your search term.

7. Leave Edge open if you want to move on to the next topic.

Search results (and advertisements)

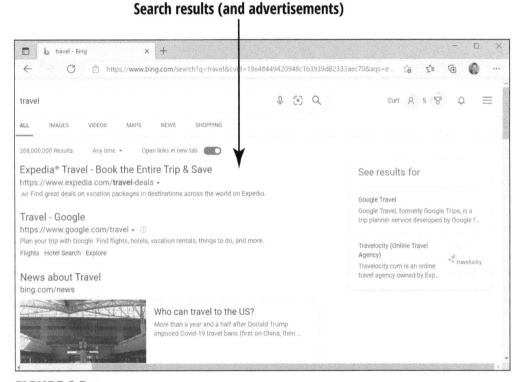

FIGURE 6-5

Bookmark and Revisit Your Favorite Websites

1. Open Edge, if it isn't already open, and go to your favorite website on the Internet (or if you don't have a personal favorite, just go to www. dummies.com). If you've spent any time on the Internet, you soon find websites that you want to visit again and again. Rather than memorize the addresses of these websites, you can add them to your Favorites list to make revisiting them quite easy.

2. Select the Add This Page to Favorites button (located on the right side of the address bar; it's shaped like a star). The Add to Favorites panel opens, as shown in **Figure 6-6.** Use this panel to describe the websites you want to revisit, and store their names in the Favorites panel (you'll take a look at the Favorites panel in Step 4).

3. Before selecting the Add button to add your favorite website to the Favorites list, consider doing the following:

   • Enter a shorter, more concise, more descriptive name in the Name text box.

   • Open the Save In menu and select a folder name to store the website in a folder. (Step 8 explains how to create folders of your own.)

4. To verify that the website you are currently visiting has been added to the Favorites list, select the Favorites button (or press Ctrl+I). The Favorites panel opens, as shown in **Figure 6-7.** Look for your website in the Favorites panel.

5. Select the Back button to go to the website you visited previously. This button is located in the upper-left corner of the screen. Next, select the Favorites button (or press Ctrl+I) and, in the list of bookmarked websites, select the website you bookmarked in Step 3 (see **Figure 6-7**). Your favorite website opens on the screen.

TIP

Don't hesitate to bookmark a website that you expect to revisit. Unless you bookmark it and add it to your Favorites list, you might not be able to find it again.

TIP

To remove a website from the Favorites list, display the Favorites list, right-click the website's name to display the context menu, and select Delete.

6. You can make the Favorites bar appear below the address bar, as shown in **Figure 6-8.** Merely by clicking or tapping a website name on the Favorites bar, you can go straight to a favorite website without having to open the Favorites panel. To display the Favorites bar, open the Favorites panel and select the See More button (refer to **Figure 6-7**). On the menu that appears, choose Show Favorites Bar Any website you really, really like is a candidate for the Favorites bar.

7. Click anywhere onscreen except in the Favorites panel to close the Favorites panel. Then select the Favorites button to open the

Favorites panel (refer to **Figure 6-7**). From here, you can manage websites that you have deemed your favorites:

- **Reorder the websites and folders:** Drag a website or folder up or down in the Favorites panel until it lands in the right place.

- **Remove a website (or folder):** Display the folder or website's context menu (right-click or touch it with your finger). Then choose Delete.

- **Create a folder:** Choose Add Folder on the Favorites panel toolbar and type the new folder's name.

- **Rename a website (or folder):** Display the folder's context menu (right-click it or touch it with your finger), choose Rename, and enter a new name.

TIP

Folders can be a big help in organizing and finding bookmarks in the Favorites panel. If you're the type who likes to bookmark websites, give some thought to creating folders to store your bookmarks.

8. Leave Edge open if you want to move on to the next topic.

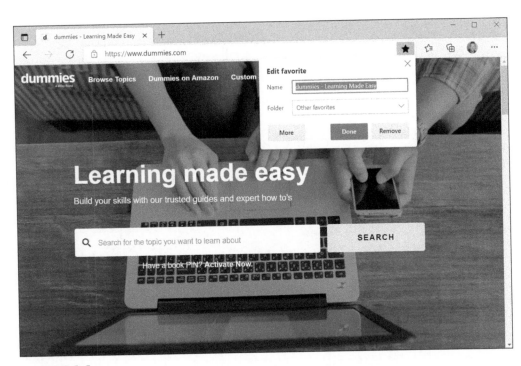

FIGURE 6-6

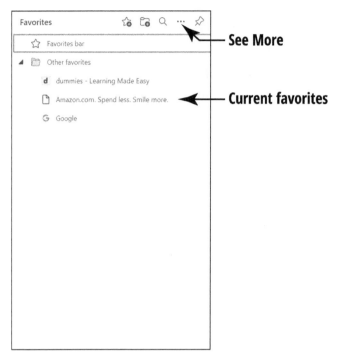

See More

Current favorites

FIGURE 6-7

Favorites bar

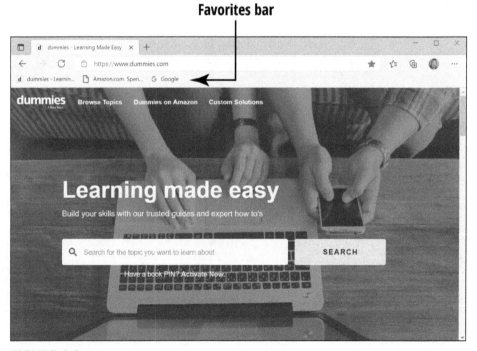

FIGURE 6-8

Explore Edge Features

1. You can decide what web pages you want to see on Edge's start screen. Go to your absolute favorite web page so that you can pin this web page to the Start screen. Select the Settings and More button (it's located in the upper-right corner of the screen). Then, on the drop-down menu, choose More Tools and choose Pin to Start on the submenu. Select Pin in the confirmation dialog box that asks whether you really want to pin this web page to the Start menu or cancel. You get a second confirmation dialog box making sure, so click Yes to that box as well (yes, it's a bit redundant!).

2. Close Edge and then go to the Start screen and select your favorite web page's tile. By selecting this tile, you do two things at one time — you start Edge and open it to your favorite website. Isn't that convenient?

3. Select the Settings and More button, and then select History on the menu that appears. As shown in **Figure 6-9,** the History panel opens. This panel offers the means to backtrack to websites you visited in the past hour, the past day, or the previous week.

4. Scroll to the bottom of the History panel to delve into ancient history and see websites you visited some time ago. Select a website to open it in Edge.

TIP

To erase your browsing history and remove all web page names, select More Options on the History panel and select Clear Browsing Data.

5. Steps 3 and 4 demonstrate how Edge keeps track of web pages you visited, but what if you resent this kind of snooping? You can browse the web without letting Edge keep a historical record of the websites you visited. To browse in private, select the Settings and More button (it shows three dots and is located in the upper-right corner of the screen) and choose New InPrivate Window in the drop-down menu. Edge opens a second, "InPrivate" window, as shown in **Figure 6-10.** Websites you visit privately are not recorded in the History panel. Close the InPrivate window for now (or browse a while and then close it).

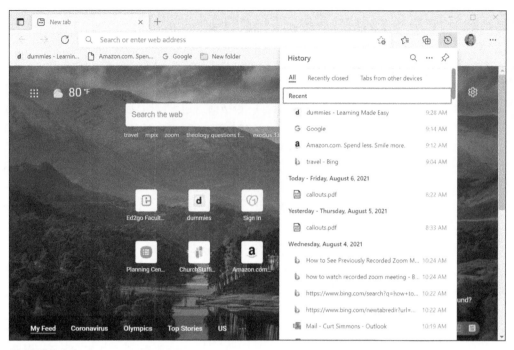

FIGURE 6-9

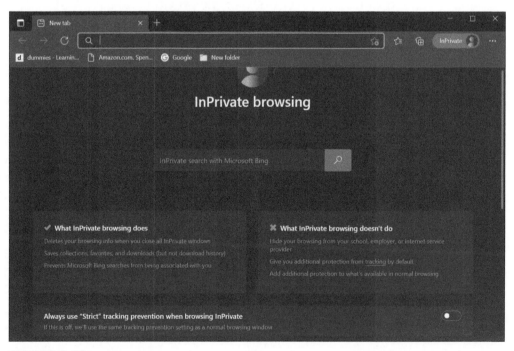

FIGURE 6-10

You can tell when you're browsing privately because the word *InPrivate* appears in the upper-left corner of the Edge window (see **Figure 6-10**).

InPrivate browsing doesn't track your browsing history or associate searches with your profile; however, it does not hide your browsing from your school, employer, or Internet service provider.

6. Have you ever found interesting things on the web, such as a photo, a recipe, or even a whole web page that you would like to save for later? You can do that with an Edge feature called Collections. Collections is a fun and easy to way to save anything you find on the web so that you can use it later. To get started, click the Collections button on the Edge toolbar and then click Next in the Collections panel, shown in **Figure 6-11.** You see a few screens with some basic info for you; just click Next through those screens.

Collections button

Click Next to get started

FIGURE 6-11

7. On the New Collection panel, you can add the current web page or even drag an image to the panel to begin a collection. In **Figure 6-12,** you can see a couple of web pages and image in a collection.

Can't seem to drag an image to the Collections panel? Just right-click the image and select Add to Collections.

If you want to remove something from a Collection, right-click the item in the Collections panel and click Delete.

8. The Collection panel has an Add Note button, which is a handy feature. For example, if you're shopping for a new car and you save several website and images, you can add your personal notes to the Collection so that you can review them later. Having a note can be a helpful way to remind you of specific details about images or websites that you don't want to forget.

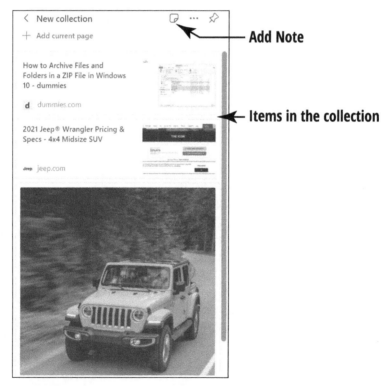

Add Note

Items in the collection

FIGURE 6-12

9. You can create as many collections as you want. Select the Collections button and then select Start New Collection on the Collections panel. For example, you could create a collection of images and websites called Birthday Ideas, or maybe one called Recipes (if you like to cook), or Vacation Ideas. How you organize your Collections is completely up to you!

10. Close Edge or keep reading if you're inclined to find out how to turn off the ads on web pages.

Block Ads on Web Pages

1. As you soon find out if you spend any time on the web, web pages are full of cumbersome advertisements. They often get in the way. How would you like to strip the ads from the web pages you visit? You can do that by installing an app from the Microsoft Store called Adblock Plus. Chapter 10 explains how to install an app from the Microsoft Store. Close Edge, go to the Microsoft Store, and install Adblock Plus now (and visit Chapter 10, if necessary). It's free, and installing takes only half a minute. We'll wait for you here.

2. The first time you open Edge after Adblock Plus is installed, you see the message screen confirming the installation. You can close this message and then go to a web page of your choice.

3. **Figure 6-13** shows the same web page before and after Adblock Plus was installed. Notice the large ad at the top of the first Edge screen. This ad is absent from the same screen on the bottom of the figure. After you install Adblock Plus, the Adblock Plus button (with the letters ABP) appears on the toolbar (refer to **Figure 6-13**). A number next to the button tells you how many ads were blocked on the web page you're visiting.

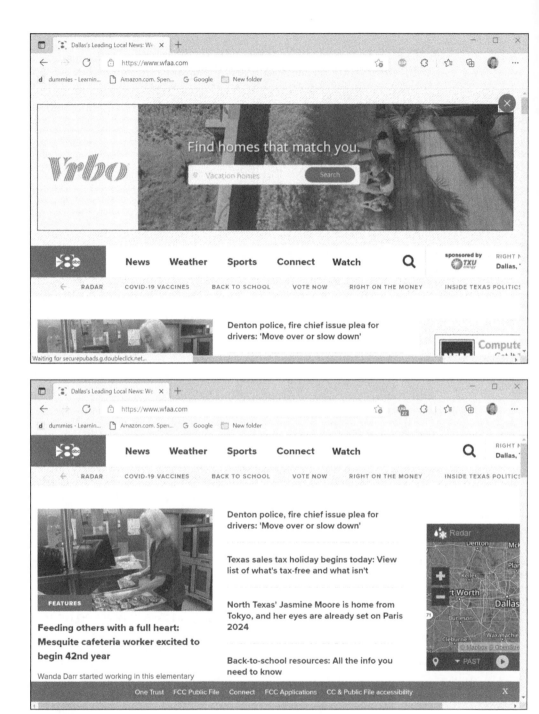

FIGURE 6-13

4. Click the ABP button on the toolbar. You see the drop-down menu shown in **Figure 6-14.** It also tells you how many ads were blocked. Occasionally you run into a web page that can't be viewed when ads are blocked. In cases like that, you can turn off Adblock Plus. To do so, click the APB button, drag the slider on the drop-down menu to the left, and select the Refresh button.

5. Click or tap anywhere on the web page to close the Adblock Plus drop-down menu.

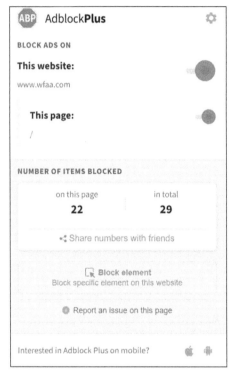

FIGURE 6-14

Although ad blocking software and InPrivate browsing can help you, know that some websites may not function properly or refuse to work altogether when you're using these tools. This doesn't happen too often, but often enough to mention!

Chapter **7**

Emailing Family and Friends

E very day, billions of email messages circle the globe, conveying greetings, news, jokes, even condolences.

Email also provides a way to send and receive *attachments*, such as documents or photos. Who needs faxes or postcards?

The Mail app gives you access to email using your Microsoft Account. Your Microsoft Account can be associated with an email address from any email service provider, including Microsoft's own Hotmail (www. hotmail.com), Live (www.live.com), Outlook (www.outlook.com), or Xbox (www.xbox.com). You can also use email services such as Yahoo! (www.yahoo.com) and Gmail (www.gmail.com) without having a Microsoft Account. You can easily set up these email services to work in the Windows 11 Mail app, too, and this chapter shows you how.

In this chapter, you use the Mail app for email. You compose, format, and send email, files, and pictures. You discover how to respond to incoming email. You also configure email accounts, and create and use email folders.

TIP

If you don't have a physical keyboard, see Chapter 1 for information on using the virtual keyboard.

Use the Mail App

1. For email on Windows 11, select the Mail icon on the taskbar. If you don't see the Mail app on the taskbar, select the Search icon on the taskbar and type **mail** in the Search box. Select the Mail app from the search results. Either way, the Mail app opens. If you're currently signed in to your Microsoft Account, the Mail app knows your email address. (You gave it to Microsoft when you created the account.) This address is listed on the Mail screen, and you're ready to go. If you're not currently signed in using a Microsoft Account, enter your email address and password on the Sign In screen that appears and then select Sign In. (See Chapter 4 for information on creating a Microsoft Account.)

TIP

The first time you use the Mail app, you are invited to "add an account." In other words, Windows 11 offers its help linking an email account you have to the Mail app. Select the Add Account button and follow the onscreen instructions for making an email account available to the Mail app. See "Add an Email Account to Mail," later in this chapter, for more information.

2. When you start the Mail app, you land in the Inbox folder. If you haven't used your Microsoft Account for email, you probably don't have any messages in your Inbox folder. Who would have written you already? You may, however, see a message or two from Microsoft.

3. On the Inbox screen, select the Expand/Collapse button, as shown in **Figure 7-1.** The Folders panel expands, and you can see the names of all your folders in the Folders panel. In the Mail app, email is kept in folders.

TIP

The number beside each folder name tells you how many unread email messages are in each folder.

4. Select any of the following folders in the Folders panel and then select the Expand/Collapse button to return to the Mail panel (you have to select More in the Folders panel to see some of these folders):

- **Outbox:** Email you have sent or is in the process of being sent.

- **Inbox:** Email you have received but not moved elsewhere appears here, including both read and unread email.

- **Archive:** The Mail app provides this folder for email that you want to keep on hand for archival purposes.

- **Drafts:** As you compose email (see next section), you can save your message as a draft until you are ready to send it.

- **Sent:** Copies of the email you send are stored here, appropriately enough.

- **Deleted:** Messages you have deleted from your Inbox. You can move messages out of the Deleted folder if you change your mind.

- **Conversation History:** Conversations you have had with people through email. For example, if you and a friend email each other back and forth, the Conversation History folder will keep track of the entire conversation, which is much easier than keeping track of individual emails.

- **Junk:** Messages are moved here — either automatically or by you — instead of the Inbox, if they are suspected of being, well, *junk.* The more common term for unwanted email is *spam.* (It's a long story.) You may want to check this folder occasionally to see whether a message was misfiled here.

You may see other folders in the Folders panel besides the folders described here. If you linked the Mail app to an email account with other folders, those folders are also listed in the Folders panel.

5. Proceed to the next section, "Write an Email Message."

TIP

To search for stray email messages, go to the Search box and enter the content you're searching for. For example, enter a subject, sender, or something in the message.

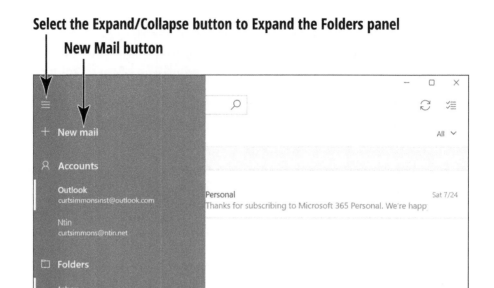

Select the Expand/Collapse button to Expand the Folders panel

New Mail button

Select More to see all folders

FIGURE 7-1

Write an Email Message

1. To compose an email message in the Mail app, select the New Mail button (it looks like a plus sign) in the upper-left corner of the screen (refer to **Figure 7-1**). (It doesn't matter which folder you're in when you select this button.) The Compose screen appears, as shown in **Figure 7-2.**

2. In the To box, type an email address. If the Mail app recognizes the address because you've sent email to or received email from the person to whom you're addressing your message, a pop-up box appears so that you can select the person's address without having to type it.

Press the Tab key, touch the screen, or use the mouse to move to the next option.

TIP

To send email to more than one person at one time, select the Cc & Bcc button (located to the right of the address line). Then type a recipient's address in the Cc box (and the Bcc box as well, if you want to send a blind copy without revealing you've done so). (Once upon a time, *CC* stood for *carbon copy,* and then it became *courtesy copy.*) Technically, it makes no difference whether you use additional To addresses or the CC.

3. Enter a few words in the Subject box describing your message's content or purpose.

4. Select below the Subject line. Type your message. Avoid all caps; purists consider caps to be SHOUTING! No need to press Enter as you approach the end of the line — this isn't a typewriter.

TIP

Some people start a message with a salutation, such as *Hi Peter,* but many people do not. Some people sign email using their initials. Email can be as formal or casual as you choose to make it.

TIP

Words underlined with a red squiggle aren't recognized by Windows 11 and may be misspelled. Right-click or tap and hold to see a pop-up menu, and choose Proofing on the menu. You see a list of suggested spellings. Choose the correct spelling if it is on the list. You can also go to the Options tab and select the Spelling button there.

5. When you're ready to send the message, select the Send button (you'll find it in the upper-right corner of the Compose window; refer to **Figure 7-2**). If you decide not to send the message, select the Discard button.

TIP

Email you compose but haven't sent yet is kept in the Drafts folder. You can postpone writing a message and finish it later by going to the Drafts folder and opening your message there. To go the Drafts folder, select the Expand/Collapse button (if necessary) and then select Drafts in the All Folders panel.

6. If you select the Send button, your message is sent and you return to the folder you were in at the beginning of Step 1.

7. Repeat from Step 1 as needed.

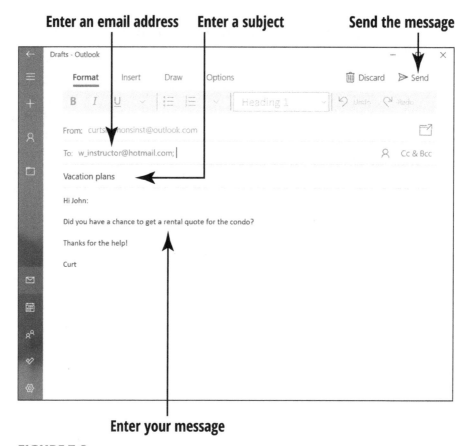

Enter an email address   **Enter a subject**   **Send the message**

Enter your message

FIGURE 7-2

TIP

Mail puts a signature at the bottom of email messages: "Sent from Mail for Windows 11." If you don't like this signature or want to put your own signature at the bottom of all the emails you send, select the Settings button on the bottom of the Folders panel. The Settings panel opens. Select Signature and either turn off signatures or enter a signature of your own.

Format Email

1. If you want to add bold, italics, or other formats to email, select the text you want to fancy up in the Compose window, as shown in **Figure 7-3.** If no text is selected, your formatting will apply to the word in which the *cursor* (the blinking vertical line in the content) is located.

- **Mouse:** Click and drag the mouse pointer over the text you want to select.

- **Touchscreen:** Tap and drag over the text you want to select.

- **Keyboard:** With the cursor at the beginning of the text you want to select, hold down the Shift key as you press the right or down arrow to select text. Release the Shift key only after you have completed your selection.

To apply formatting to one word, double-click or double-tap that word to select it and display the app bar in one step.

The keyboard shortcut to select all text is Ctrl+A. No mouse or touchscreen method is quite so complete.

2. Using the commands on the Format tab, select from the following formatting options:

- **Bold:** Bold is used for emphasis to make text stand out.

- **Italic:** Although italic can also be used for emphasis, it may be harder to read than normal or bold text.

- **Underline:** Because links are usually underlined automatically, you may want to avoid underlining text that isn't a link.

- **Bulleted list:** Select the Bullets button and enter items for the list. As you enter items, Mail attaches a bullet point to the start of each line.

- **Numbered list:** Select the Numbering button and type the list. As you type, Mail assigns a number to each line.

- **Styles:** Choose a style from the drop-down menu. A style applies many formats simultaneously. Heading 1, for example, enlarges the text and turns it blue.

See Chapter 1 for information on using the virtual keyboard's emoticons.

TIP

To create a hyperlink, go to the Insert tab and select the Link button. Then enter the address destination of the hyperlink in the address bar that appears. See Chapter 6 for more information about hyperlinks.

3. When you finish formatting the text, select the Send button.

Format the text

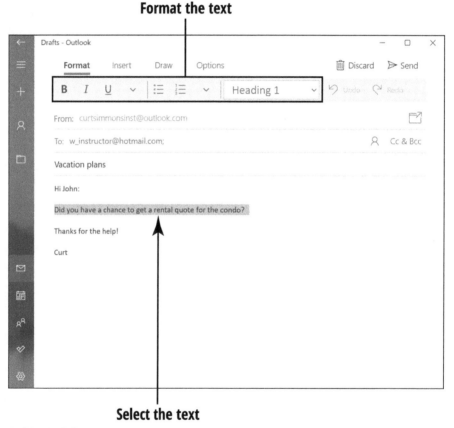

Select the text

FIGURE 7-3

Send Files and Pictures

1. To send a file or picture with your email message, go to the Insert tab. As shown in **Figure 7-4,** the Insert tab offers tools for sending files and pictures.

2. Select the Files button to send a file. The Open dialog box appears. Locate and select the file you want to send and then select the Open button in the dialog box. You can send more than one file with an email message. In the Compose window, files you want to send appear as thumbnails under the heading "Attachments" (see **Figure 7-4**). Click the X button on a file if you change your mind about sending it.

3. To send a picture with an email message, place the cursor where you want the picture to be. Then select the Pictures button on the Insert tab. The Open dialog box appears. Select the picture and then click or tap the Insert button.

TIP

You can format pictures before sending them. Click or tap a picture to select it in the Compose window. The Picture tab appears. Using the tools on this tab, you can rotate, crop, and resize pictures.

4. When you have finished attaching your files and pictures, select the Send button to send them on their merry way.

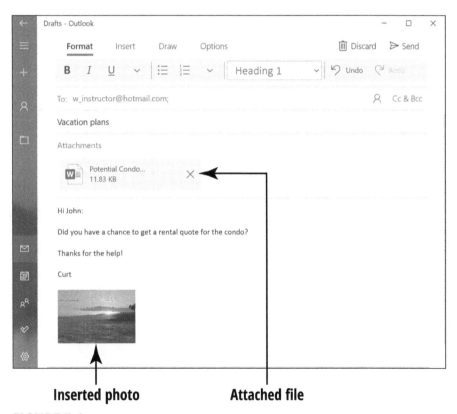

Inserted photo   **Attached file**

FIGURE 7-4

TIP

Chapter 11 explains how you can use the Photos app to scan photos to send them to others.

Read and Respond to Incoming Email

1. On the Start screen, select the Mail tile. The Mail app opens to the Inbox folder.

2. Select a message under the Inbox heading. The Inbox view changes to the message view, as shown in **Figure 7-5.**

TIP

In email terminology, bouncing messages back and forth with someone about the same subject is called a *conversation.* When you select an email message that is part of a conversation, the other messages in the conversation appear as well so that you can select and reread them. You can also find these conversations in the Conversation History folder.

3. After you read the message, you can respond to or forward it if you want. Select the Reply or Reply All button to reply; select the Forward button to forward the message. Selecting any of these buttons starts a new message (refer to **Figure 7-2**) that includes the text of the original message. The subject line is automatically *RE: [the original subject]* (in the case of replies) or *FW: [the original subject]* (in the case of forwarded messages). Complete your message and select the Send button.

TIP

Reply All sends your response to all the other recipients, if a message is addressed to more than one person. The Reply option sends your response only to the sender. Select Forward to send the selected message to someone else. You can add your own text or remove portions of the forwarded message in the process.

TIP

See the sections "Write an Email Message" and "Format Email," earlier in this chapter, for more information on composing an email message.

4. Note the following options for managing messages (to take advantage of some of these options, select the More button in the upper-right corner of the screen):

- **Delete:** Use this option to delete the message.

- **Set Flag:** Use this option to place a flag icon next to the message in the Inbox so that you remember to deal with the message later on. You can swipe right to flag a message.

- **Mark Unread/Read:** Use this option if you want the message to appear unread (marked with a horizontal blue line in the Inbox). Select the More button (the three dots on the right side of the toolbar) to see this option. Some people mark a message as unread when they want to deal with it later.

- **Move:** Use this option to move the selected email from one folder to another. To access the Move option, select the More button (refer to **Figure 7-5**). After you select Move on the More drop-down menu, select a folder in the Move To panel to move your email.

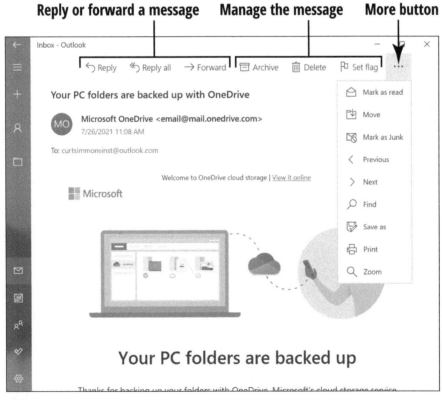

FIGURE 7-5

TIP If you add email accounts from providers such as Gmail, you may see additional categories or options. See the section "Add an Email Account to Mail" at the end of the chapter.

Change Mail Settings

1. In Mail, display the Settings panel, shown in **Figure 7-6.** To display the Settings panel, display the Folder panel and select the Settings button (you'll find it in the lower-right corner of the Folder panel). The Settings panel appears on the right side of the Mail screen.

2. On the Accounts panel, select Manage Accounts, and then select the email account with settings that need a change. The Account Settings dialog box appears (see **Figure 7-7, left**).

3. Make any desired changes on the Account Settings dialog box and select the Save button:

- **Account Name:** The name of the email service provider. You can change this, if you want. Curt might use *Curt's email.*

- **Change Mailbox Sync Settings:** Click or tap here to open the Sync Settings dialog box (see **Figure 7-7, right**) and change how and when the Mail app collects email. Skip to Step 5 if you choose this option.

- **Delete Account:** Deletes the email account. The Mail app is no longer associated with the account after you delete it.

4. In the Sync Settings dialog box (refer to **Figure 7-7, right**), choose settings and select the Done button:

- **Download New Content:** If you don't want email constantly streaming into your Inbox, you can change this setting to every 15 or 30 minutes, Hourly, or only when you select the Sync This View button (which is located next to the Search button).

- **Always Download Full Message and Internet Images:** Select this option if pictures aren't appearing in your incoming email messages.

- **Download Email From:** You can limit how far back to download messages to your computer. This setting isn't relevant in a new email account.

- **Sync Options:** Turn the Email setting to Off only if you don't want the Mail app to receive email from the account. Turn the other two settings, Calendar and Contacts, to Off if — for the purpose of privacy — you don't want the Mail app to retrieve information from the Calendar app or People app.

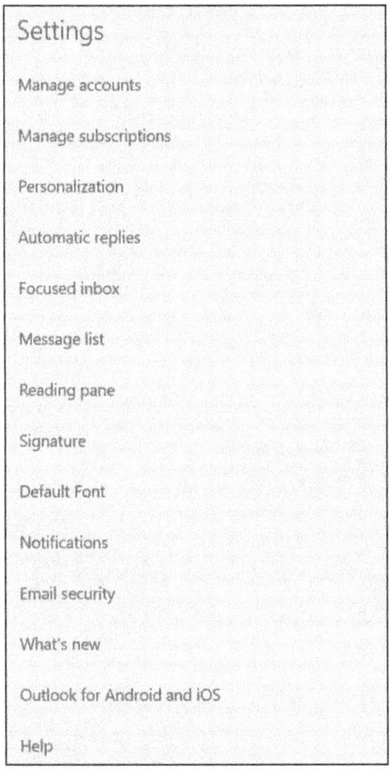

Settings
Manage accounts
Manage subscriptions
Personalization
Automatic replies
Focused inbox
Message list
Reading pane
Signature
Default Font
Notifications
Email security
What's new
Outlook for Android and iOS
Help

FIGURE 7-6

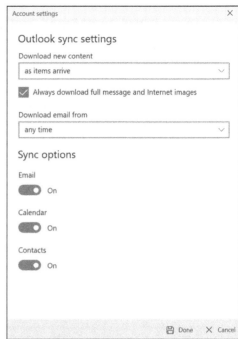

FIGURE 7-7

Add an Email Account to Mail

1. If you have another email address, you can add it to Mail. Adding an account has definite advantages. Instead of going hither and yon to collect your email from different accounts, you can collect all your mail in one place — the Mail app. The following steps explain how to add an existing Gmail account to Mail. In Mail, display the Settings panel (refer to **Figure 7-6**) and then select Manage Accounts.

2. In the Manage Accounts panel, select Add Account.

TIP

Many people have more than one email address. Your Internet service provider probably gave you an email account, and you may have another through work or school. However, this feature isn't for everyone.

3. In the Add an Account dialog box, select the service with which you have an email account. If your service doesn't appear here, select Other Account. If you have a Gmail address (or you just want to see what's next), select Google.

4. On the Connecting to a Service screen, enter your Gmail address, select Next, and enter your password in the appropriate box. Then select Next. Select the Back button if you don't want to continue.

5. The next screen informs you how your Gmail account will connect with the Mail app. Scroll to the bottom of this screen and select the Allow button.

6. In the dialog box that says you've added an account, select Go to Inbox.

7. Return to the Manage Accounts folder (select the Settings button and choose Manage Accounts in the Settings panel). As shown in **Figure 7-8,** the Manage Accounts panel shows you the names of your accounts. You can switch to a different account by selecting its name in the Manage Accounts panel.

FIGURE 7-8

See the preceding section, "Change Mail Settings," for the steps to review or change settings for your newly added account.

Create Email Folders

You may have a certain person with whom you exchange emails often, such as a friend, adult child, or even grandchild. If you want to keep these messages, you can create a custom folder just for that purpose. This way, you can easily access those emails any time you want.

1. In Mail, click the All Folders button to display the All Folders panel.

2. Select the Create New Folder button, shown in **Figure 7-9.**

3. Enter the new folder name in the dialog box that appears and press Enter. The new folder now appears in the All Folders list.

Organize Email in Folders

1. All email you receive initially appears in Mail's Inbox. You can move email that you want to keep to an appropriate folder. (You can create as many folders as you like, as shown in the previous section.) To move an email, first select the email in the Inbox to open it.

2. Select the Actions menu and then select Move. You can also right-click an email in your Inbox and select Move.

3. In the All Folders list that appears, select the folder you want to move the email to. The email message is moved to that folder, and you can access it any time by selecting that folder.

TIP

If you move an email message to a folder but later want to move it to a different folder, you can easily do so. Just open the folder where the message is currently stored and repeat the previous steps to move it to a different folder.

All Folders button   **Create New Folder**

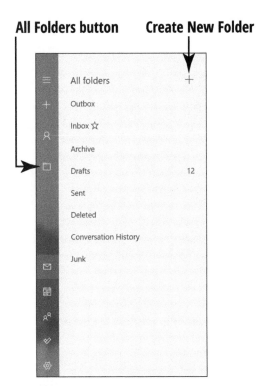

FIGURE 7-9

Chapter **8**

Connecting with Microsoft Teams

Keeping in touch with family and friends is one of the most important things you can do. Windows 11 wants to make this easy for you with some new software called Microsoft Teams that is built into the operating system. Teams is, in a nutshell, communications software. Using Teams, you can easily chat with people you know, make video calls, and even schedule meetings.

Although the Teams software scales up to big, enterprise organizations where thousands of people in different locations can connect and work together, this chapter focuses on the Teams features that are most helpful to you.

TIP

You can also connect with people using Skype (see Chapter 9). Skype is also included with Windows 11, but Microsoft is phasing out Skype and it will be gone at some point. Teams is a better way to go now.

You can also download the Teams app and use the mobile app on your phone. That way, you can always stay in touch with people whether you're using your computer or your phone. The Teams app is free — just search for it on your phone's app store.

TIP

Use Chat

1. On the taskbar, click the Chat button. The Chat flyout panel appears. The Chat flyout panel is a part of Teams, but Windows 11 gives you the flyout panel as a quick and easy way to chat or meet with someone, as shown in **Figure 8-1.**

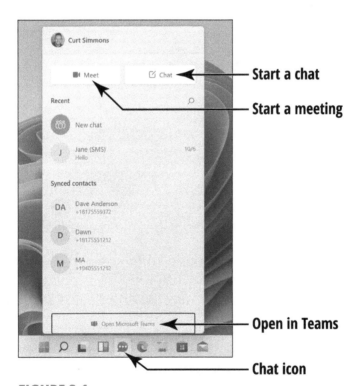

Start a chat

Start a meeting

Open in Teams

Chat icon

FIGURE 8-1

2. To start a chat, click the Chat button (see **Figure 8-1**). The New Chat window appears, shown in **Figure 8-2.** Teams attempts to locate contacts you already have on your computer (such as old Skype contacts, Mail app contacts, and so forth). But you can directly reach a contact with an email address and phone number as well. In the New Chat window, enter the name of an existing contact, or use an email address or phone number.

If you want to chat with multiple people at the same time, repeat Step 2 and then click Add a Group Name. For example, you could chat with your children or grandchildren all at one time, and name the group. This way, you easily chat with all the group members in the future. Note that if you want to chat with two or more people at the same time, the New Chat window will require you to add a group name. (See **Figure 8-2**.)

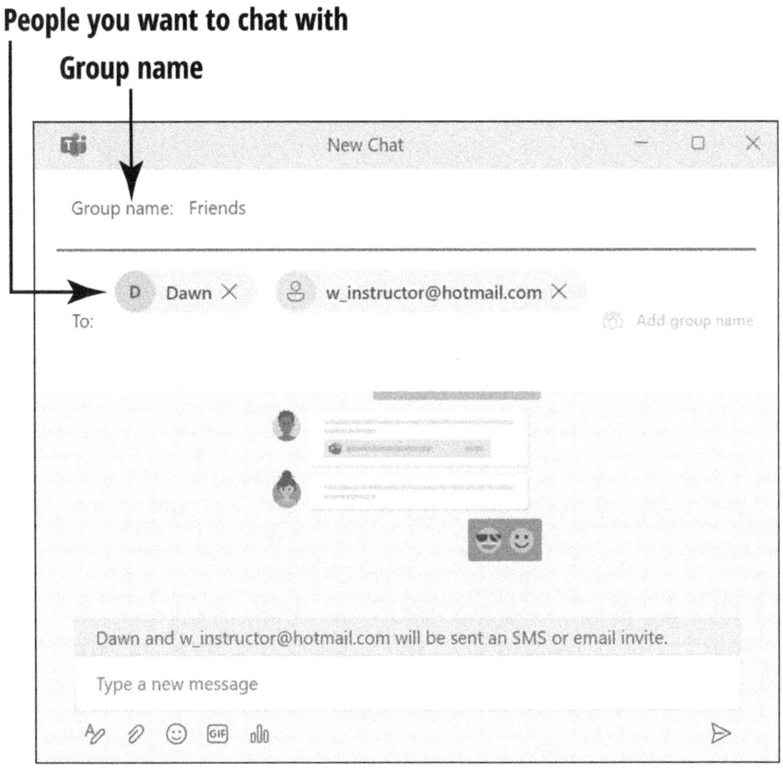

People you want to chat with

Group name

	New Chat	— □ ✕

Group name: Friends

To: D Dawn ✕   w_instructor@hotmail.com ✕   Add group name

😎 🙂

Dawn and w_instructor@hotmail.com will be sent an SMS or email invite.

Type a new message

FIGURE 8-2

If you want to chat with multiple people, you can use a contact name, an email address, or phone number — in other words, you can mix and match these different contact methods.

3. Type the message you want and select the Send button, shown in **Figure 8-3.** Your contacts are sent your message.

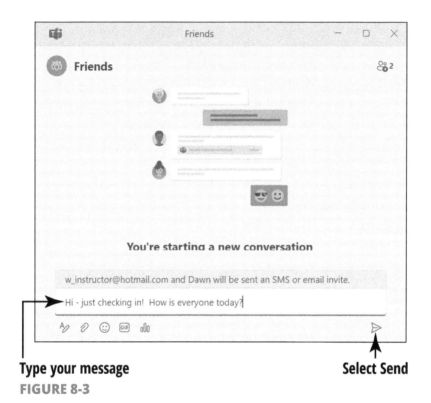

Type your message   **Select Send**
FIGURE 8-3

4. As you chat, you can see what you've written as well as see responses from others. You can simply chat back and forth, as shown in **Figure 8-4.**

5. Note that you can attach files as well as send emojis and GIF images if you like. Just click those button options under the Chat message box to use them (see **Figure 8-4**).

TIP

If you attach emojis, GIFs, or images, they appear as a part of your chat rather than as attachments, as shown in **Figure 8-5**. This is a fun way to share, especially photos!

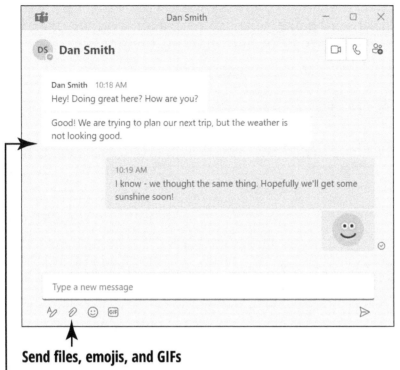

Send files, emojis, and GIFs

Chat conversation

FIGURE 8-4

FIGURE 8-5

TIP

In the Chat flyout panel, you also have a Meet option, which starts a video call. Selecting this button opens the full version of Teams.

Chat in Teams

1. On the Chat flyout panel, select the Open Microsoft Teams link (refer to **Figure 8-1**). You can also start Teams any time using these methods:

   - Clicking the Search icon on the taskbar and searching for Teams.

   - Clicking the Microsoft Teams icon in the system tray of the taskbar (next to the date and time). You may need to select the Show Hidden Icons button to see the Teams icon.

2. The Teams interface gives you an app bar, a navigation pane, and then a content pane (which is similar to other windows in Windows 11, such as File Explorer). For example, in **Figure 8-6,** Chat is selected on the toolbar. Previous chat conversations appear in the navigation pane, and then the selected chat conversation appears in the content pane.

3. Select Chat on the toolbar. You can respond to any previous chat conversations by selecting the conversation in the navigation pane. Then just start chatting in the content pane. You can also start a new chat by selecting the New Chat button (refer to **Figure 8-6**).

4. When you select New Chat, the new chat appears in the contents pane. Enter the name, email, or phone number of the person you want to contact and then type your message, shown **in Figure 8-7.** This works the same way as chatting from the taskbar chat icon.

TIP

Chatting in Teams works the same as chatting from the Chat flyout panel available on the taskbar. The advantage of using Teams is that you may find it easier to jump from conversation to conversation because Teams gives you a navigation pane and a content pane in the same interface.

Toolbar   **New chat**

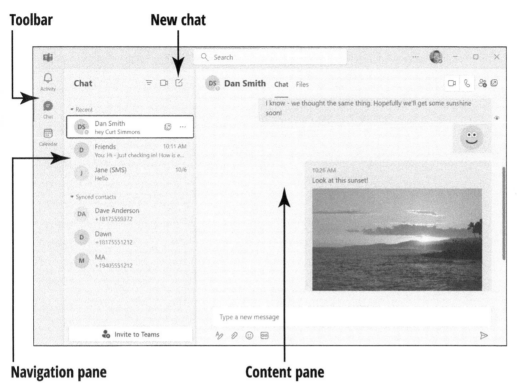

Navigation pane   **Content pane**

FIGURE 8-6

TIP

If you're chatting with someone, you can easily jump to a video or audio call by clicking the button options (refer to Figure 8-7). See the "Start a Meeting" section, later in this chapter.

Respond to Activity

1. In Teams, select Activity on the toolbar. The Activity section shows you specific mentions (Teams calls them @mentions) of you. For example, you may be chatting with five people in a group. When someone specifically mentions you, that mention appears in Activity, as shown in **Figure 8-8.** This feature helps make sure you don't miss anything specific to you.

Chat conversation

Pop-out chat button

Video or Audio call

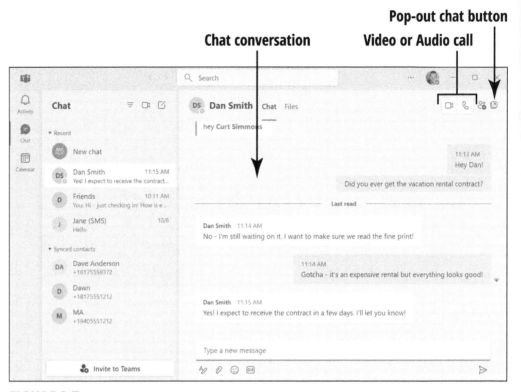

FIGURE 8-7

TIP

When someone mentions you, or you mention someone, you type *@username*. For example, if you have a contact named Sue Williams and you want to mention Sue, you type **@Sue Williams**. Your mention then shows up in Sue's Activity in Teams. Note that a mention has an @ symbol with it and is pronounced "at mention."

2. With Activity selected on the toolbar, click the @mention you want to see in the Feed (navigation pane). Then you can respond to the @mention on the content pane (see **Figure 8-8**).

3. @mentions appear in the Feed as bold type, meaning that you haven't read them yet. You can mark an @mention as read after you read it or if you don't want to read it. Just point to the @mention and select More Options to the right. You see the option to mark the @mention as read, shown in **Figure 8-9.**

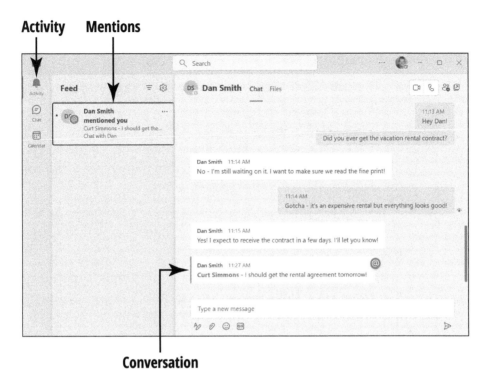

Activity   **Mentions**

Conversation

FIGURE 8-8

FIGURE 8-9

Manage Chats

1. In Teams, select Chat on the toolbar.

2. Choose a chat conversation in the navigation pane and select More Options (point your mouse to the chat to see the More Options button appear; it's the three dots).

3. The More Options menu gives you a few different ways to manage chat conversations, as shown in **Figure 8-10:**

- **Pop-Out Chat:** This option opens the chat in a new, pop-out window.

- **Mark as Read / Unread:** You can mark a chat as either read or unread. If you've read the chat, it will default to "read," but you can change it back to unread if you want (which is a good way to remind yourself to respond to chat later).

- **Pin:** This option pins the chat conversation to the top of your Chat list in the navigation pane. Doing so makes it easy to return to conversations you're actively engaged in.

- **Mute:** Say you're chatting with a group of friends, but the conversation spirals into details that don't pertain to you. You can mute the conversation so that you don't get updates as people continue to chat.

- **Block:** This option blocks a chat and the people chatting with you. Make sure you want to use this feature before you block someone from getting in touch.

- **Hide:** This option hides the chat from the navigation pane so that you won't see it any longer.

- **Delete:** Finished with a chat? You can choose to delete it from the navigation pane. Be careful: After you delete a chat conversation, it's gone forever!

4. You can also filter your chat conversations. For example, if you have a lot of conversations in the chat navigation pane, it can be difficult to find the one you want. You can select the Filter option on the navigation pane and see only the chats from a specific person, as shown in **Figure 8-10.** This will make finding specific chat conversations easier.

TIP

You can invite your friends and family to join Teams so that you can more easily connect with them. At the bottom of the navigation pane, select the Invite to Teams button. A window appears with a link, as shown in **Figure 8-11.** Click Copy. You can then paste this link into email messages, text messages, and so forth. The link gives your friends and family a quick way to download the Teams software and get connected with you. A Microsoft Account is required for anyone using Teams.

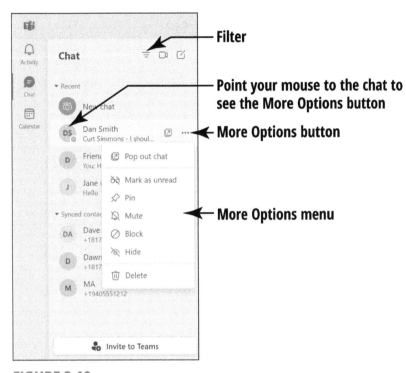

FIGURE 8-10

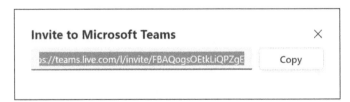

FIGURE 8-11

Start a Meeting

1. In Teams, a *meeting* is a live video call. To start a meeting, select Chat on the toolbar and select the Meet Now button, shown in **Figure 8-12.**

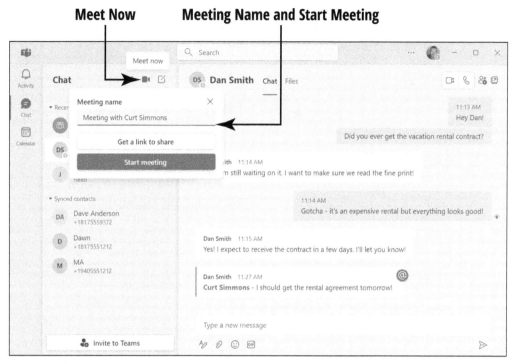

Meet Now   **Meeting Name and Start Meeting**

FIGURE 8-12

2. In the dialog box that appears, give the meeting a name (you can accept the default name). You can then get a link for the meeting, which you can send to others whom you want to attend, or you can start the meeting. Click Start Meeting.

3. The Teams interface changes to a video meeting window, shown in **Figure 8-13.** Before you join the meeting, there are a few important settings you can set up here:

- **Turn on Audio:** This setting is on by default; otherwise, your microphone will not work during the meeting.

- **Turn on Video:** This setting is on by default; otherwise, your webcam will not work during the meeting.

- **Background Filter:** This option, shown in **Figure 8-14,** blurs the background of your video. This option is helpful to reduce clutter and increase visibility in your video. It also helps provide some level of privacy (unless you want everyone to see the unfolded laundry behind you!).

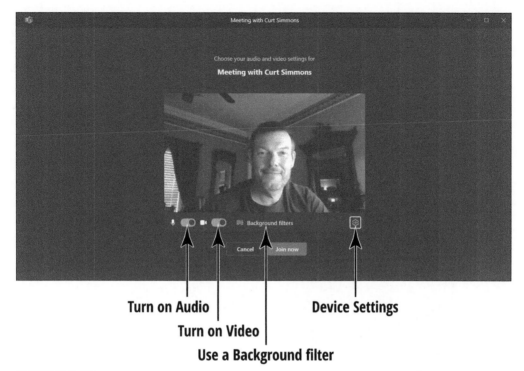

Turn on Audio

Turn on Video

Device Settings

Use a Background filter

FIGURE 8-13

Background appears blurry

Blur the Background Filter

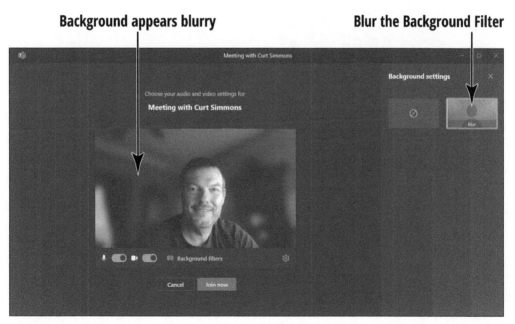

FIGURE 8-14

- **Device Settings:** This feature opens a panel where you can select the speakers, microphone, and camera you want to use for the video call, shown in **Figure 8-15.** This feature is really only helpful if you have multiple microphones or webcams connected to your computer. This way, you can choose the ones you want to use.

4. Select Join Now. The Video window opens. You see yourself as well as any other participants as they join the meeting. Now you can visit virtually with your friends, family, and any other person or group you want to meet with.

FIGURE 8-15

When you're finished meeting, just select the red Leave button in the upper-right corner of the screen.

TIP

Schedule a Meeting

1. In Teams, select Calendar on the toolbar and choose the New Meeting button, shown in **Figure 8-16.** The odds are good that you'll want to schedule meetings with other people. When you schedule a meeting, you create a date and time and then send a meeting link to the people you want to invite so that you can all meet at the same time. This is a great way to schedule time with others.

Choose Calendar   **Select New Meeting**

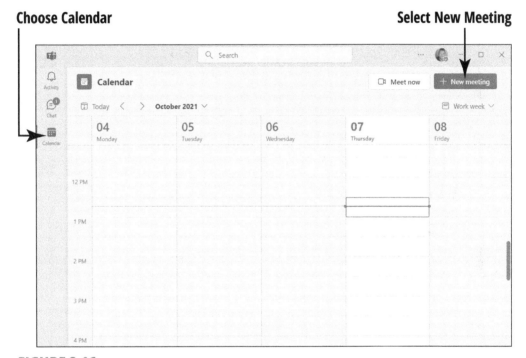

FIGURE 8-16

2. In the New Meeting window, make sure that your time zone is correct, and then enter the details of the meeting, as shown in **Figure 8-17.** When you're done, select Save.

3. Teams creates the meeting and gives you a dialog box from which you can copy the meeting link and even add it to Google Calendar. Either way, you want to share the meeting with the link or calendar with the people you want to attend.

Select Save when you're done

Enter the meeting details

FIGURE 8-17

TIP

When you send a meeting link, all the participant has to do is click the link to open the meeting in Microsoft Teams.

TIP

Need to cancel a meeting? No problem — select Calendar on the toolbar and locate your meeting on the calendar. Right-click the meeting and select Cancel. You can also right-click the meeting and choose Edit to change the date, time, and other details.

Manage Teams Settings

1. In Teams, select Settings and More and choose Settings from the menu, as shown in **Figure 8-18.**

2. In the Settings pane, choose Notifications. Teams will notify you if someone is trying to connect with you through notifications and sounds. You can manage the notifications and sounds by clicking

Open Windows Notifications Settings. Here, you can determine what notifications you receive. For more information about Windows 11 notifications, see Chapter 3.

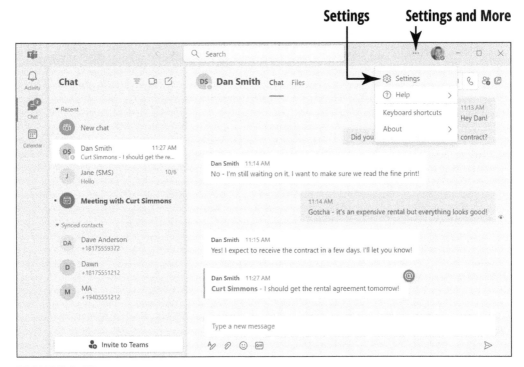

FIGURE 8-18

3. Select Appearance and accessibility in the Settings panel. Here, you can choose a theme for Teams. By default, Light is selected, but you can choose Dark or High Contrast if you like, as shown in **Figure 8-19.**

4. Select Privacy in the Settings panel. The default settings here are all you need, but there is one setting you should note. By default, when you read a chat or anything else in Teams, the sender can see that you've read it (called a read receipt). You can turn this feature off with the toggle button if you don't want others knowing whether you've read their messages.

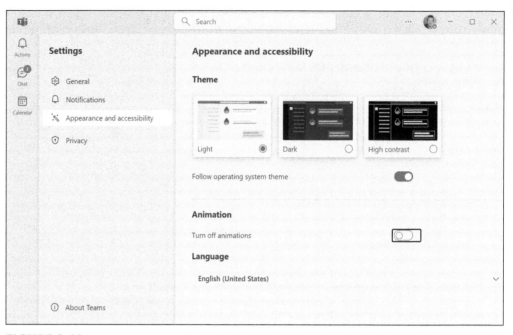

FIGURE 8-19

Chapter **9**

Exploring Apps for Your Daily Life

This chapter describes a handful of apps that the average Joe or Jane might use every day. It shows how to manage tasks and lists to keep track of your life with the To Do app. You also use the Calendar app to keep track of birthdays, anniversaries, appointments, and other events.

This chapter delves into the Maps app, showing you how to use it to get directions and to look up hotels and restaurants when you're on the go. You see how two Windows 11 apps, Sticky Notes and Alarms & Clock, can help you complete tasks on time, and how WordPad can help you with your word processing chores.

Finally, this chapter demonstrates how to make the Lock screen indicate whether you've received email, how to chat face-to-face with family and friends over the Internet with Skype, and how to ask questions of Cortana.

Add a Task with Microsoft To Do

1. Select the Search icon on the taskbar and search for To Do. Select the To Do app that appears in the search results. The To Do app opens and likely opens with the Getting Started option selected in the left pane. If not, select Getting Started, shown in **Figure 9-1.**

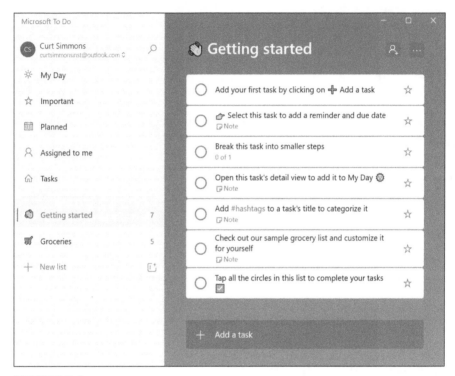

FIGURE 9-1

2. Select Add a Task in the right pane. Type the name of the task. For example, you could type **Pay phone bill**, as shown in **Figure 9-2.**

3. After you type a name, three options appear in the name field: Add a Due Date, Remind Me, and Repeat (refer to **Figure 9-2**). Select Add a Due Date, shown in **Figure 9-3,** and select one of the options, or choose pick a date to choose a date from a calendar.

4. To set a reminder, select the Remind Me option and choose a default reminder time to appear, or you can pick a date and time. When you select a reminder, Windows 11 displays a reminder on your computer on the designated date and time.

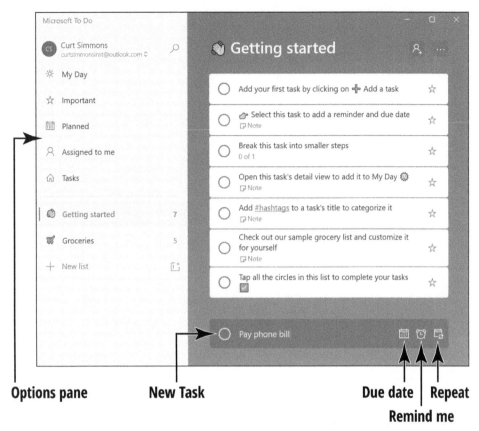

Options pane   **New Task**   **Due date** | **Repeat**

Remind me

FIGURE 9-2

5. If you want to want to make the task recurring, select the Repeat option. This feature is handy for tasks you do every week or month, such as paying a bill. The task will automatically recur, and if you set up a reminder, you'll also get an automatic reminder as well. When you select Repeat, just choose how often you want to repeat the task (daily, weekly, and so on), or you can create a custom repeat schedule.

6. When you're done, click the plus sign (+) that appears next to the task to save it (see **Figure 9-4**).

7. Tasks that have a due date appear in the Planned section in the left pane of the app. If you select Planned, you see a lists of tasks that you've created and when they're due, as shown in **Figure 9-5.**

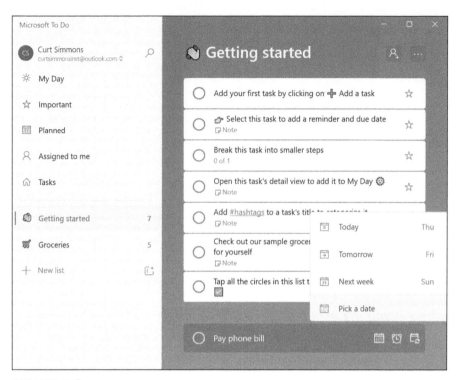

FIGURE 9-3

FIGURE 9-4

TIP

When you finish a task (or decide you don't want it any longer), just select the circle next to the task's name. This will make the task show as completed.

Edit a Task

1. Sometimes thing change and you need to edit a task as well. For example, say you have an appointment that gets moved to a different day or time. You don't need to create a new task — you can just edit

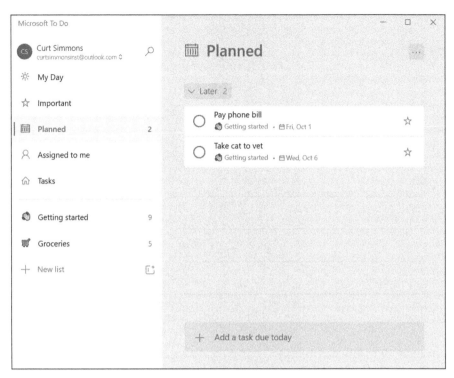

FIGURE 9-5

the existing one quickly and easily. Open the To Do app and locate the task you want to edit (you may find it by selecting Planned or Tasks in the left pane).

2. Select the task you want to edit. A pop-up panel appears with options to edit the task. For example, in **Figure 9-6,** Curt has an appointment to take the cat to the vet. He can simply select the date to change it and save the new date.

3. You can also change the Remind Me schedule or the Repeat schedule on this pane. Just select those options to edit them.

4. Change your mind and want to delete a task completely? No problem — just right-click the task and select Delete Task.

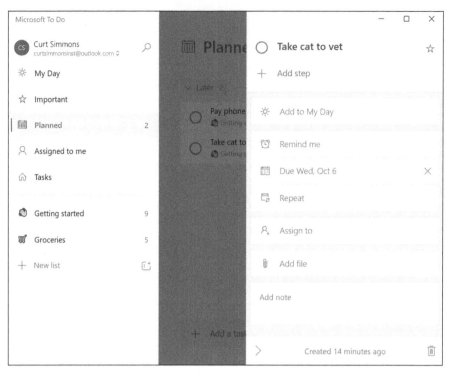

FIGURE 9-6

Create a List

Individual tasks are helpful, but sometimes you need a group of tasks that revolve around the same thing. In this case, you want to create a To Do List. For example, say you're planning a birthday party for a family member and have several ideas you don't want to forget. A list can help you remember them!

1. Open the To Do app and in the left pane, select New List. Type a name for the list and press Enter. The new list appears, as shown in **Figure 9-7.** Here, we've created a list to plan a birthday party.

2. Now, begin adding tasks to the list by clicking the Add a Task option at the bottom of the app. You can create as many tasks as you need for your list. The tasks appear under the list name, as shown in **Figure 9-8.**

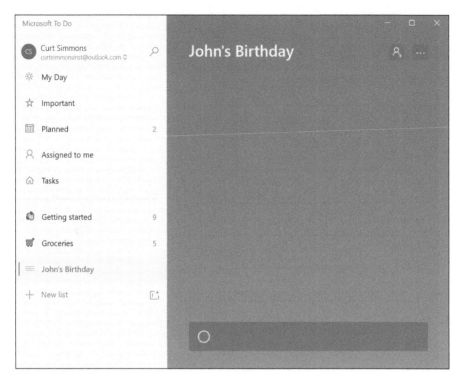

FIGURE 9-7

3. You can make changes to any task in a list as needed. You can complete a task, edit it, or even delete it if you want. Refer to the previous "Edit a Task" section.

TIP

You can do other things with the To Do app, such as plan your day, flag important tasks, and manage groceries, but everything in the app revolves around Tasks and Lists, so feel free to explore the app a bit more. Also, you can download the Microsoft To Do app on your phone, and it will sync with your computer's To Do List automatically (if you're signed in with your Microsoft Account). Just search for Microsoft To Do at your app store to find and download it. The mobile app is a great tool when you need a list of to-do items at your fingertips or reminders about groceries you need — after all, you don't want to forget the ice cream!

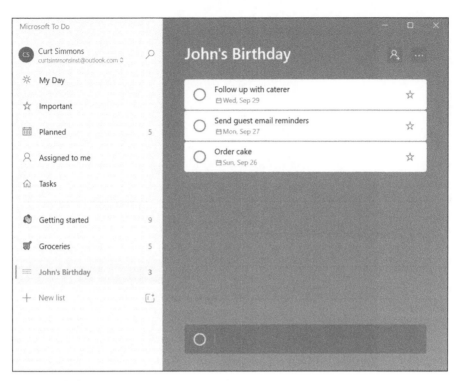

FIGURE 9-8

Add a Birthday or Anniversary to the Calendar

1. To see the calendar and add events, open the Calendar app by selecting the Search icon on the taskbar, typing **calendar** in the Search box, and selecting Calendar in the search results. You may see events and appointments from services you connected to in Mail.

2. On the Calendar app bar located along the top of the screen (refer to **Figure 9-9**), select each of the following formats:

   • **Day:** This format displays a single day with a box for each hour. Select arrows near the top of the screen to scroll back or forward one day at a time. Scroll up and down to see more hours in the day.

   • **Week:** The current week appears. Scroll or use the arrows to go from hour to hour or week to week.

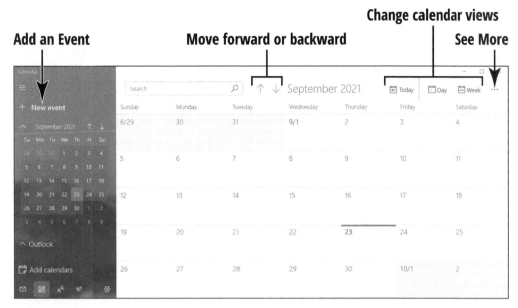

Add an Event
Move forward or backward
Change calendar views
See More

FIGURE 9-9

TIP

Open the drop-down menu on the Week button and choose Work Week to see the current workweek (Monday through Friday).

- **Month:** The current month appears in the classic month layout. (Depending on the size of your screen, you may have to scroll to see this format.) Today is highlighted with a different color. Use arrows near the top to move forward and back a month at a time.

- **Year:** The months of the year appear. You can select a month to open a month in Month view.

TIP

Select the Today button on the Calendar app bar to go immediately to today's date on the Calendar.

TIP

If you don't see the Month or Year option, select the See More button to view them.

3. The Calendar offers two ways to add an event: the short-but-cursory way and the slow-but-thorough way:

- **Short but cursory:** Click or tap the day on which the event is to occur (in the Month format) or the hour and date on which it is to occur (in the Day or Week formats). You see the pop-up window shown in **Figure 9-10.** With luck, this little window is all you need to describe the event. Type the event's name, and if it isn't an all-day event, deselect the All Day check box (if necessary) and use the start time and end time menus to describe when the event starts and ends. You can also type the event's location. Select the Save button when you finish describing the event (or select the More Details link to open the Details screen and go to Step 4).

- **Slow but thorough:** Select the New Event button (refer to **Figure 9-9**). The Details screen appears, as shown in **Figure 9-11**. Move on to Step 4.

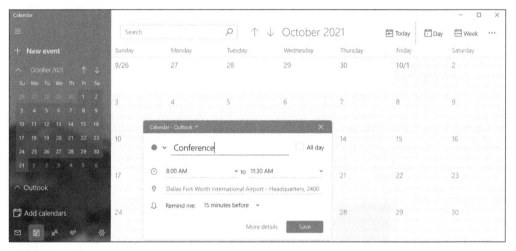

FIGURE 9-10

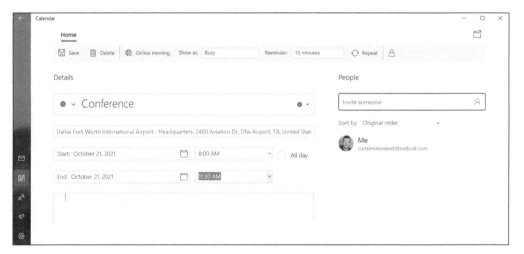

FIGURE 9-11

4. Under Details, enter or (or change) any of the following data for the event:

- **Event Name:** Describe the event in one to three words. The description you enter will appear on your calendar.

- **Location:** If listing the location will help you get to the event on time, by all means list it. The location, like the event name, appears on the calendar.

- **Start and End:** Select the calendar icon, and on the pop-up menu that appears, choose the day on which the event begins and then the day on which it ends.

- **Hours/All Day:** If this isn't an all-day event, deselect the All Day check box, if necessary. Then use the drop-down menus to describe when the event will occur.

TIP

If the event is one that recurs, you can schedule it on a weekly, monthly, or yearly basis on the calendar. Weekly staff meetings, birthdays, and anniversaries are examples of recurring events. Select the Repeat button on the app bar along the top of the screen to schedule a recurring event. Then, using the Repeat options that appear in the Details screen, describe how often the event occurs, when it occurs, and when it will cease occurring.

- **Event Description:** Describe the event in detail if you think it's necessary. Should you wear the tux or the zoot suit? The ball gown or the formal dress? This is the place to record your thoughts on the matter.

5. Optionally, use the commands on the app bar at the top of the screen to further describe the event:

- **Show As:** Choose an option on this menu if you share your calendar with someone, such as a receptionist, who may schedule you for other events. You can declare yourself free, tentative, busy, or out of the office.

- **Reminder:** How far in advance of the event do you want Calendar to display a notification? Choose None, 5 Minutes, 15 Minutes, 30 Minutes, 1 Hour, 12 Hours, 1 Day, or 1 Week.

- **Private:** Select this option (look for the lock symbol) to prevent this event from appearing on a shared or public calendar. The event remains visible on your own screen.

6. When you're done, select the Save button. (Select the Delete button if you don't want to create an event.)

7. On the calendar, your event appears on the specified date and time.

8. To edit an event, select it on the calendar. A pop-up screen appears with details about the event. Select the event's name in the pop-up screen to open the Details screen, where you can add or change any detail.

To delete an event, open it in the Details screen and select the Delete button (a trash can).

If your Microsoft Account is connected to a third-party service such as Gmail, events from that service appear on your calendar. To load third-party service information into the Calendar app, select the Settings button (in the lower-left corner of the screen) and select Manage Accounts in the Settings panel. You see a list of accounts. Select the account, and in the Account Settings screen, select Change Mailbox Sync Settings. Then, on the Sync Settings screen, choose whether to sync the calendar on the third-party service with the calendar you see in the Calendar screen.

Search and Explore with the Maps App

1. Use the Search icon on the taskbar and search for Maps; then select the Maps app in the search results. Use this app to locate places and to get driving or walking directions.

2. Click or tap in the Search box. A drop-down list appears. It lists items you already searched for, if you searched already. You can select an item on the list to revisit it.

3. Type **1 california st, san francisco** (you don't need to enter capital letters) in the Search box. As you type, options appear in the Search panel. Select the address you entered, not the names of businesses

located at the address. As shown in **Figure 9-12,** the Maps app shows you where the address is located; it also provides a photo. As well as typing addresses in the Search box, you can enter the names of places and landmarks.

Enter an address or search term   **Directions button**

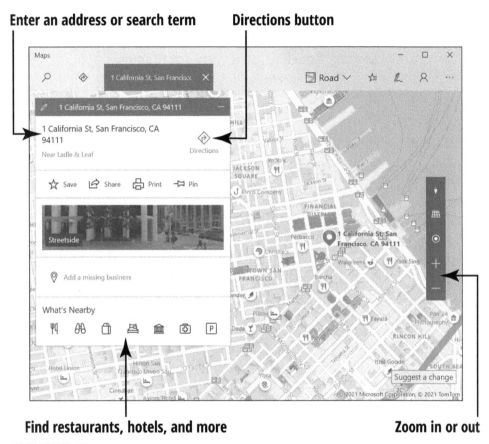

Find restaurants, hotels, and more   **Zoom in or out**

FIGURE 9-12

4. Select the Zoom Out button (or press Ctrl+minus sign) on the app bar in the map. Then select the Zoom In button (or press Ctrl+plus sign). The tools on the app bar can help you read the map better (refer to **Figure 9-12**).

5. Select the Directions button (refer to **Figure 9-12**), which is the button in the upper-right corner near the Search box. (Or you can press Ctrl+D.) Tools appear to help you get from place to place. Click or tap in the Starting Point box (Box A), and when the drop-down menu

appears, choose **1 California St,** the address you already entered. In the Destination box (Box B), enter **1501 haight st** and choose the 1501 Haight St address in San Francisco in the drop-down menu that appears.

6. Select the **Get Directions** link. As shown in **Figure 9-13,** the Maps app gives you several sets of driving directions between the two addresses and tells you how long each drive will take. Select a drive in the list to get specific instructions for driving between the two addresses. After you select a drive, you can click or tap the Print icon to print the instructions.

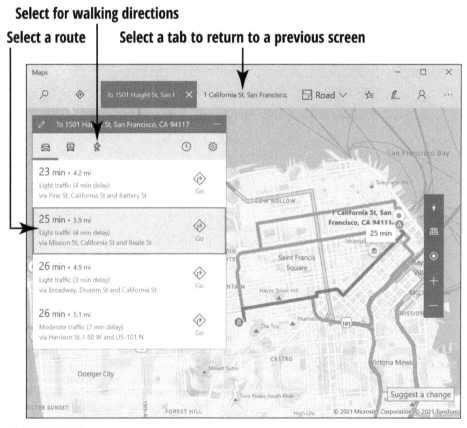

FIGURE 9-13

7. Are you walking, not driving? In that case, select the Walking icon at the top of the Search panel to get directions for walking between the two locations. You can also select the Transit icon to get directions for getting there by public transportation.

8. Select the Search button (the magnifying glass in the upper-left corner of the screen) and then select **1501 Haight St., San Francisco** in the drop-down menu. You get a detailed look at the streets surrounding this address. Select the Restaurants icon under What's Nearby (refer to **Figure 9-12**) to see restaurants on the map. You can select the name of a restaurant to find out more about it. The What's Nearby icons also provide the means to find attractions, parking, hotels, and other goodies.

9. The Maps app keeps tabs on your travels. Select a tab along the top of the screen to retrace your steps and revisit a map you saw earlier.

Stay on Time with the Clock App

1. For the most part, the Clock app is a simple tool that gives you a few different options to set timers. Open the Clock app by searching for Clock, using the Search icon on the taskbar. You see a navigation pane on the left side of the app where you can toggle between different options (see **Figure 9-14**). They are:

   - **Focus Sessions:** You can choose Start Focus Session to create a timer for something specific. For example, if you're exercising on a treadmill and you want to walk for a specific time, you can create daily focus sessions and track your progress over time (refer to **Figure 9-14**).

   - **Timer:** You can activate some present timers (1 minute, 3 minutes, 5 minutes, and 10 minutes). You can also click the + icon to create a new timer. When you start a timer, a chime will sound when the timer is complete.

   - **Alarm:** You can set a daily alarm (or choose which days you want the alarm to sound). For example, you could set a daily wake-up alarm, a daily exercise alarm, and so forth.

Alarm

Timer

Focus Sessions

Navigation pane

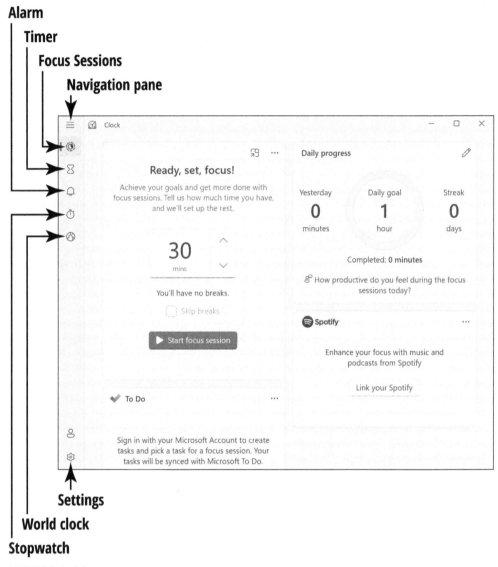

Settings

World clock

Stopwatch

FIGURE 9-14

- **Timer:** Here you find a simple timer you can use. You can also select the Lap/Split button while you're using the timer to mark laps and such. This feature is helpful if you're exercising and want to flag the time a lap took to complete.

- **World Clock:** This lets you view the local time, but you can also select the + button to see the time and date in any city in the world.

2. You can select the Settings option (refer to **Figure 9-14**) to make some adjustments to the Clock app. For example you can turn on or off end-of-session or break sounds (and adjust the sounds as well). You can also adjust the app theme if you like.

Write Documents with WordPad

1. Windows 11 comes with a word processing app called WordPad. Word processors are applications that you can use to write letters, reports, and other documents. WordPad isn't as sophisticated as its cousin Microsoft Word, the world's most famous word processor, but it's pretty darn good. And it doesn't cost anything. Moreover, you can open documents created in Word in WordPad (though the document may not display all of Word's formatting correctly). Word can read WordPad documents, too. To start WordPad, type **WordPad** using the Search icon on the taskbar and then select WordPad in the search results. WordPad opens, as shown in **Figure 9-15.**

FIGURE 9-15

2. Type a few words and then select the words by dragging over them. You can tell when words are selected because they are highlighted. After you select words, you can reformat them. Open the Font Family menu on the toolbar and choose your favorite font. Then open the Font Size menu and choose a font size. By changing fonts and font sizes, you can change the words' appearance.

3. Click or tap the Bold button (the B) on the Home tab (or press Ctrl+B). The words are boldfaced. Boldfacing, italicizing (select the Italic button), and underlining (select the Underline button) are other ways to change the appearance of words.

4. Select the Linespace button and choose an option — 1.0, 1.15, 1.5, or 2 — on the drop-down list that appears. In this way, you can change the amount of space between lines. For example, choosing 2 double-spaces the text you selected.

TIP

You can choose formatting commands *before* you start typing. The commands you choose are applied to the words you write next. It isn't necessary to write the words first and then select them to make formatting changes.

5. Drag the Zoom slider in the lower-right corner of the screen to the right (or click or tap the Zoom In button on the right side of the slider). Zooming in increases the size of text on the screen so that you can see it better. Drag the slider to the left (or click or tap the Zoom Out button) to make the text look smaller.

6. Press Ctrl+S or select the Save button (located in the upper-left corner of the screen) to save your document. You see the Save As dialog box. WordPad documents are saved in the RTF format (RTF stands for *rich text format*). As we mention earlier, Microsoft Word can read and display RTF files. Enter a name for your document and select the Save button.

TIP

Chapter 15 explains how to open, save, and otherwise manage files.

Write a Sticky Note to Yourself

1. Use the Search icon and search for Sticky Notes. Select the app in the search results to open it. A sticky note — like the ones in **Figure 9-16** — appears on the screen. Sticky notes are the digital equivalents of the yellow stick-'em tabs that people attach to their computers and desks. Use sticky notes to remind yourself to do something or be somewhere.

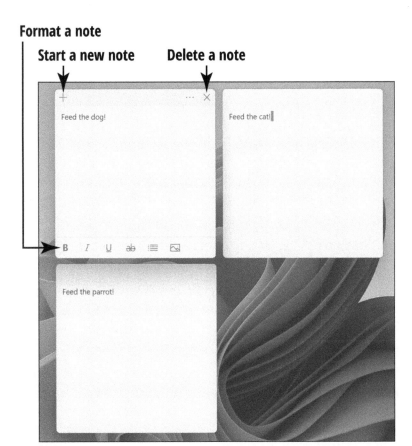

Format a note

Start a new note   **Delete a note**

Feed the dog!

Feed the cat!

B I U ab ≡ ⊠

Feed the parrot!

FIGURE 9-16

2. Type the note. Then move the pointer over the stripe at the top of the note and drag the note to a corner of the screen.

TIP

You can change the size of a sticky note. Move the pointer over the lower-right corner of the note. When the pointer changes to double arrows, start dragging.

3. Click or tap the New Note button (refer to **Figure 9-16**). Another sticky note appears. Write a second note and drag it to a corner of the screen.

4. Click or tap the Delete Note button on one of your notes when you no longer need it. You can also format the note text or even add an image (refer to **Figure 9-16**).

Chat Face-to-Face with Skype

1. Select the Search icon on the taskbar and type **skype** in the Search box; then select the Skype app in the search results. You can use this app to video chat with others. (You can also use the Microsoft Teams app to video chat; see Chapter 8 for more information.) To use Skype, your computer must have audio and video capability. You can also use Skype if a webcam is connected to your computer.

TIP

You need to sign in to your Microsoft Account before you use Skype. See Chapter 4 for information about creating a Microsoft Account.

2. The first time you use the Skype app, the screen asks you to sign in with your Microsoft Account as well as add a photo (if you like), and then it asks you to test your microphone and webcam. Just follow the instructions to check everything.

3. The Skype screen lists the names of people on your Contacts list and people you have called before using Skype. You can select one of these people to call them now. If the person you want to call isn't listed, enter their name in the Search Skype box and select the Search button (which looks like a magnifying glass). The Skype directory opens, as shown in **Figure 9-17.** Scroll to find the person you want to call, using the person's photo ID and geographical information as your guide.

4. When you see the name of the person you want to call, select the name. Then, on the call screen, select the Video Call button. You hear a ring tone as the call is made, and you see the person's face

if they answer. Your face appears in the lower-right corner of the screen so that you can see what your face looks like to the caller. (You can drag this screen from place to place.) Start talking. Smile. Trade the latest news. Have a ball.

Search   **Create a new contact**

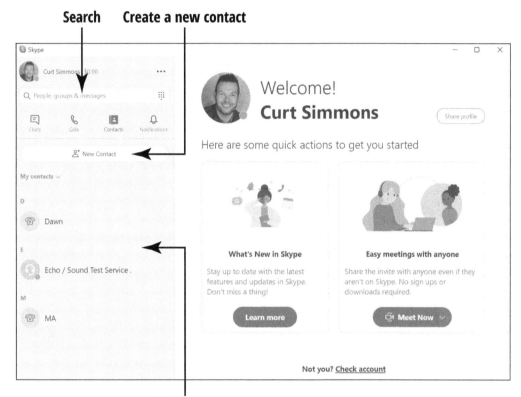

Select a name

FIGURE 9-17

TIP

Select the Video and Microphone buttons when you need a little privacy. Select the Video button to turn off your video (the other caller can't see you); select the Microphone button to turn off your audio (the other caller can't hear you).

TIP

Select the Open Conversation button (in the upper-right corner of the screen) to send an instant message to the other caller. The message panel appears. In the bottom of the message panel, type your message and press Enter. Sending text messages is a useful way to communicate when you have technical difficulties or the other caller isn't answering. Select the Close Conversation button (at the top of the Chat window) to close the message panel.

5. Select the red End Call button to complete your call.

6. When someone calls you, the Incoming Call panel appears. Select Turn Video On (at the bottom of the panel) to initiate the call.

Ask Questions of Cortana

1. Select the Search icon on the taskbar and type **cortana** in the Search box; then select the Cortana app in the search results. The first time you use Cortana, the app asks you to sign in with your Microsoft Account. Click or tap the Speak to Cortana button (it looks like a microphone) on the lower-right side of the app window. Cortana can respond to questions you ask with your voice.

TIP

To use Cortana, your computer's microphone must be up and running. Turn to Chapter 14 if the microphone on your computer isn't connected or working properly.

2. Say aloud, "Tell me a joke." As you say these words, they appear in the text box, and if all goes right, Cortana recites a joke. As well, the words of the joke appear on the screen, as shown in **Figure 9-18.**

3. Select the microphone button and ask, "What is the weather in Chicago, Illinois?" Cortana gives you the weather in Chicago.

4. Select the microphone button and ask, "What is the safest car?" As shown in **Figure 9-19,** Cortana quickly searches the Internet and provides an answer. Regardless of whether Cortana can answer a question well, you can always select the See More Results on Bing.com link in Cortana to open your browser to the Bing website and conduct a search there.

5. Cortana can remind you to attend events and appointments. Say "Remind me." Cortana replies, "Sure thing, what should I remind you about?" or "What do you want to remember?" Next, say "Walk the dog." Cortana asks when you want to be reminded. Reply, "In five minutes," and answer "yes" when Cortana asks you to confirm that you want to be reminded. As long as you begin by saying "Remind me," Cortana can record reminders like this.

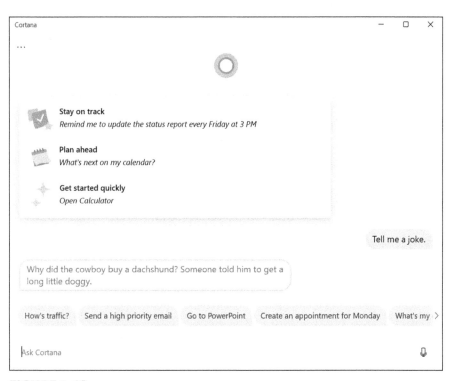

FIGURE 9-18

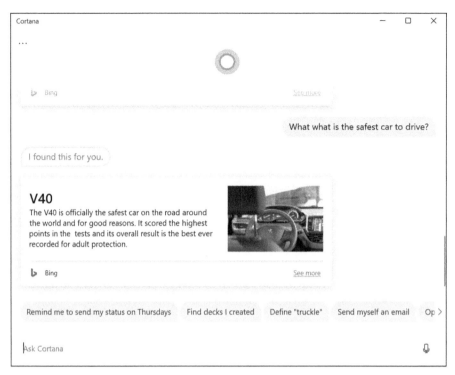

FIGURE 9-19

3

Having Fun with Windows 11

Chapter **10**

Exploring the Microsoft Store

You can do many things with Windows 11, such as send email, browse the web, play games, and look at photos and videos. You can read the news and chat with family and friends. All these functions and more involve computer programs. Back in the day, such programs were called *applications*. Now, we call them *apps*.

Windows 11 comes with a few apps installed, such as the Weather and Maps apps. (See Chapter 2 for information on using these two apps and apps in general.) To obtain other apps — free or otherwise — you use the Microsoft Store.

To install an app from the Microsoft Store, you need a Microsoft Account. See Chapter 4 for information on setting up a Microsoft Account.

Microsoft tests and approves all apps in the Microsoft Store. For quality and security purposes, you can install Windows 11 apps only from the Microsoft Store.

In this chapter, you peruse the apps and games in the Microsoft Store, including those already installed on a new machine. You install a new game, discover how to manage the apps and games you own, and see how to rate and review an app or game. Finally, you find out how to make sure that the Microsoft Store can receive payments in case you want to buy an app or game.

Explore Apps and Games by Category

1. Select the Microsoft Store icon on the taskbar. It looks like a shopping bag with the Windows 11 logo on it.

2. Look over the Store home screen, shown in **Figure 10-1.** The home screen shows top-rated apps and games, as well as apps and games that Microsoft thinks you will be interested in based on your previous purchases, if you made any. Notice the navigation bar along the left side of the screen (see **Figure 10-1**). It offers six choices: Home, Apps, Gaming, Movies & TV, Library, and Help. Wherever your travels take you in the Store, the navigation bar appears along the left side of the screen so that you can return to the home screen or redirect your search for apps, games, entertainment, and whatnot.

TIP

The Microsoft Store is, well . . . a store! Occasionally stores reorganize, and this is true for online stores as well. The good news is that you can typically figure out how to navigate and access different categories, even if Microsoft changes the look of things from time to time.

3. Select Apps on the navigation bar to go to the Apps screen.

4. If you scroll down this window, you see different app categories, such as Apps Specials, Best Selling, Productivity apps, and so forth. Select the See All button next to the Top Free Apps heading, as shown in **Figure 10-2.** Select the See All button when you want to investigate all the items listed under a particular heading. After you select this button, you see a Filters button option. You can select Filters to refine your search, as shown in **Figure 10-3.**

Navigation bar

Search

Account

Search apps, games, movies and more

Home

Apps

Gaming

Movies & TV

Home

Microsoft *Flight Simulator™*
GAME OF THE YEAR

Flight Simulator
Premium Deluxe
GOTY Edition

Top free apps

See all

Library

Help

NETFLIX

Disney+

FIGURE 10-1

5. In the Top Free drop-down menu, select Top Free to see apps that you can install for free (see **Figure 10-3**). On the All Categories drop-down menu, select Education to see educational apps. Try using the Best Selling so you can see what's most popular.

6. Select Gaming on the navigation bar and scroll to the bottom of the Games screen. You pass the familiar Best Selling and Favorite headings. You can select the See All button next to a heading to investigate all games in a heading. The Gaming screen works just like the Apps screen.

7. Return to the Store home screen by selecting the Home button on the navigation bar.

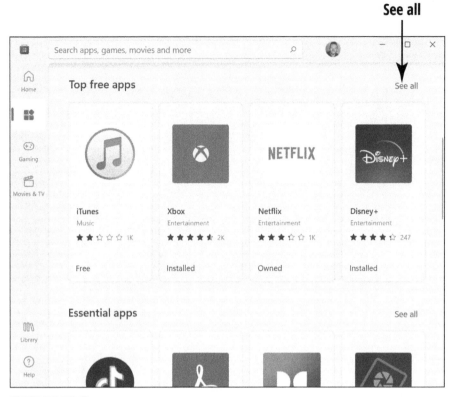

FIGURE 10-2

TIP

Did you see an app or game you want to acquire as you completed these steps? If so, skip to "Install a New App or Game," later in this chapter.

Filter options

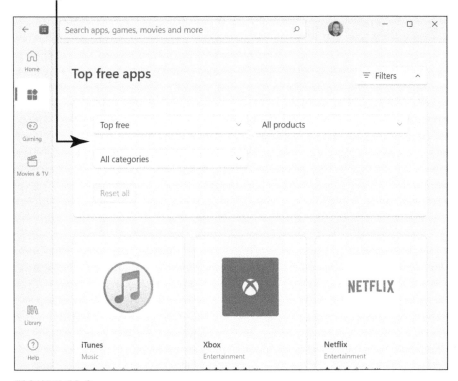

FIGURE 10-3

Search for an App or Game by Name

1. On any Store screen, click in the Search box. This box is at the top of the App Store window, as shown in **Figure 10-4.**

2. Type **chess** in the Search box and press Enter. As shown in **Figure 10-4,** games and apps with the word *chess* in their names appear in the Results screen. The Results screen tells you how many games and apps match your search. After this initial search, you can use the Filters feature to refine it if you like.

Search

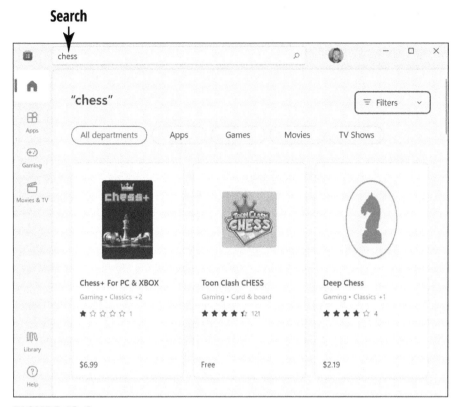

FIGURE 10-4

3. Select the Games button to see games with *chess* in their titles, as shown in **Figure 10-5.** After you select Games, you can select Filters to further refine your search if you like.

4. Proceed to the next section, "Install a New App or Game," or return to the home screen.

Games button

FIGURE 10-5

Install a New App or Game

1. In the Store, search for the Chess Free game. (See the section "Search for an App or Game by Name" for details.) You'll see several free chess games in the search results, so just choose one and select it.

2. The screen shown in **Figure 10-6** appears. This (and any) game or app screen offers lots of information.

    The Back button (left-pointing arrow) on the top left side of the screen takes you back to the previously viewed screen.

TIP

3. Notice that you see a description of the game as well as information about the number of ratings and the average. If you scroll down the page, you see more information about the game and ratings as well as a Get button.

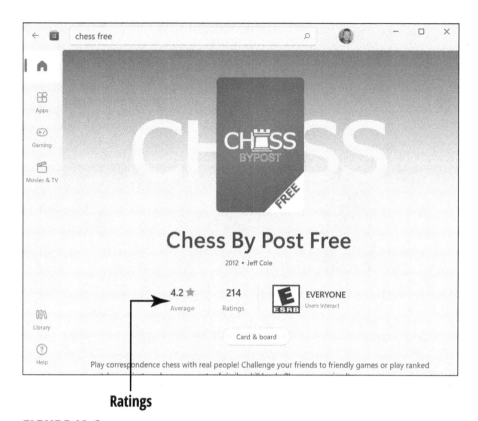

Ratings

FIGURE 10-6

4. If you scroll down the page, note the section titled System Requirements. The Microsoft Store will automatically check your system to make sure that anything you download will work on your particular computer system (most of them should).

5. Select the Get button to download and install the free chess game.

6. If you're not already signed in, enter your Microsoft Account email address and password on the screen that pops up. Then select Sign In.

TIP

Many apps are free. Before you buy an app, see the section "Add Billing Information to the Microsoft Store," later in this chapter.

TIP

If you purchase an app, you are buying a license to install that app on up to five machines using the same Microsoft Account. The Microsoft Store will track how many times and on which machine the app is installed.

7. The download indicator appears and a progress bar shows the game being downloaded and installed.

Installing an app may take a few seconds or a few minutes. You can do anything while an app installs — except use the app.

TIP

8. To play the chess game or to use any game or app after you install it, select the Search icon on the taskbar; then just type the name of the app to find it, as shown in **Figure 10-7.** Also note that you may see the newly installed app appear on the Start screen.

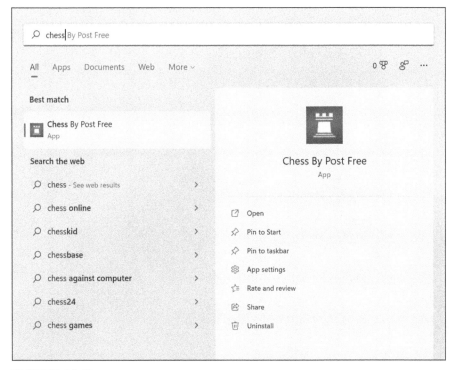

FIGURE 10-7

You can pin the chess game icon to the Start screen. See Chapter 3 for information on rearranging icons on the Start screen.

TIP

Examine Your Apps and Games

1. The Microsoft Store keeps track of which apps are installed on your Windows 11 computer. To see which apps and games you have, open the Microsoft Store and then select the Library icon on the navigation bar, as shown in **Figure 10-8.** On this screen, you can see all apps that you have downloaded, and you see button options next to each, such as Launch, Manage, Open, and so forth.

Category buttons

Installed apps

Library button

FIGURE 10-8

2. Notice the category buttons available. By default, All is selected, but you can choose to see only Apps, Games, and Movies & TV (see **Figure 10-8**). Just select the button option you want.

Rate and Review an App or Game

1. On the Library screen in the Microsoft Store (refer to **Figure 10-8**), select the free chess game you downloaded. If you don't have this game, see the section "Install a New App or Game," earlier in this chapter, or substitute any app or game you have installed.

2. In the Chess Free window, scroll down and in the Ratings and Reviews category, select the Rate and Review button, shown in **Figure 10-9**. The review screen appears (refer to **Figure 10-10**). Every app and game has a Rate and Review button.

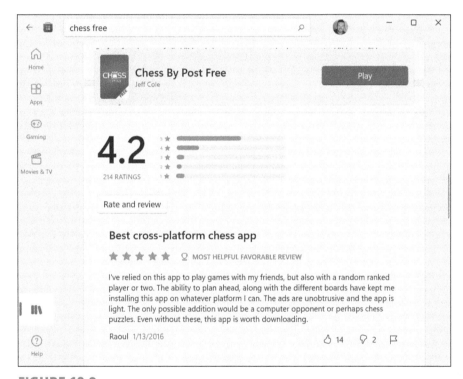

FIGURE 10-9

TIP

You can rate or review only apps and games that you've installed, and you must be signed in with your Microsoft Account.

3. In the review screen, choose the number of stars you want to give this game. (Select one star for the lowest rating; select five for the highest rating.) Select the Add a Review button.

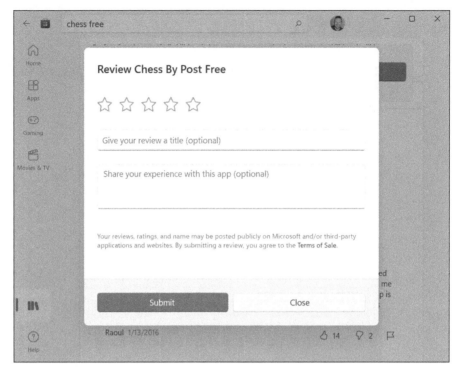

FIGURE 10-10

4. In the Write Your Review screen, say what you will, up to 500 characters, including spaces and punctuation.

5. In the Headline box, type a summary or overview of your comments, such as *Great app* or *Needs work*.

TIP

Your comments may help another person decide whether to install an app. In addition, the app's developer may use customer feedback to fix problems or add features.

TIP

Your Microsoft Account name and picture, if you have one — but not your email address — appear with your review.

6. Select the Submit button when you're done with your review.

Uninstalling an App

1. You may download an app, such as a game, and later decide you don't like or want the game any longer. You can easily uninstall anything you download from the Microsoft Store. First, search for the app you want to uninstall by using the Search icon on the taskbar.

2. Select the app in the search results.

3. In the right pane of the search results, you see the option to Uninstall the app. Select the Uninstall option, as shown in **Figure 10-11.**

4. A dialog box appears, asking you to confirm. Select Uninstall again, and the app is uninstalled from your computer.

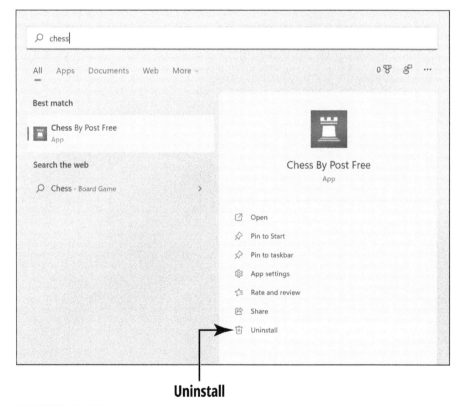

Uninstall

FIGURE 10-11

TIP

Available apps change often on the Microsoft Store. If you uninstall an app, there's no guarantee it will be available for download later. So, make sure you have no plans to use the app in the future before you uninstall.

Add Billing Information to the Microsoft Store

1. To add the billing information necessary to buy apps and games, select the Profile button next to the Search box. A pop-up menu appears with the option to select Payment Methods, as shown in **Figure 10-12.** Select this option, and a web page opens in Microsoft Edge.

TIP

See Chapter 4 for information on creating a Microsoft Account.

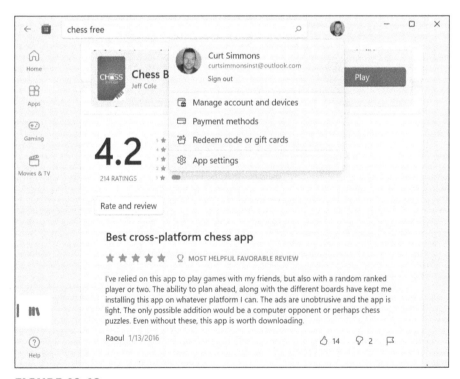

FIGURE 10-12

2. For security purposes, Microsoft asks you to enter your email address and password on the next screen. Enter your password and select the Sign In button.

3. Scroll through the screen to make sure that the information here is accurate. If you are not being billed to the correct credit card or you want to change the credit card to which you are billed, select Payment & Billing on the toolbar and choose Payment Options on the drop-down menu that appears.

4. Select a payment method (or select Payment Options if Microsoft already has credit card information on you) and do the following:

   - Select your credit card type.

     You must fill out all parts of this form except for Address Line 2.

   - Enter your credit card number in the box provided. Don't enter dashes or spaces.

   - Under Expiration Date, select the month (MM) and year (YYYY) your card expires.

   - Enter your name as it appears on your credit card.

   - Under CVV, enter the three- or four-digit verification code from your credit card. (Select What's This? for an illustration of the location of this code on your card.)

5. In the Billing Address section, do the following:

   - Enter your street address, city, state, and ZIP code.

   - Select your Country/region.

   - Under Phone Number, enter your area code in the first box and the remainder of your number in the second box.

6. When you're ready to continue, select Save.

7. If any of your data is incomplete or invalid, the form remains onscreen. Look for indications of a problem. Review each entry before selecting Submit.

8. If your information was accepted, you return to the Your Account screen. Under Payment and billing info, note your credit card type, the last four digits of your number, and the expiration date.

Chapter **11**

Taking Photos and More

Windows 11 makes enjoying digital photos easy. You can pick and choose photos to look at or display a group of photos in a slide show. You set a favorite photo to be your Lock screen background so that you see it every time you start Windows 11.

If you have a printer, you can print photos for yourself or to send them to someone. Even black-and-white prints of color photos may be nice. If you have a scanner, you can even scan photos to make digital copies of photographs that as yet exist only in paper form.

Of course, if you want to take your own photos, nothing beats having a digital camera. Copy photos from your camera to your Pictures folder for viewing and sharing. Or use the Camera app with the built-in camera found in many laptops and tablets.

In this chapter, you use the tools that come with Windows 11 for working with, printing, scanning, and editing photos.

Take Photos (and Videos) with Your Computer

1. Select the Search button on the taskbar and type **camera**. Windows 11 will locate the Camera app for you. If you don't have a webcam, the app screen displays *We can't find your camera.* If you don't have a built-in webcam, or the resolution of the one you have is too low, you can easily add a webcam. Simply plug the camera into your computer — it's that easy.

TIP

You can choose from many good webcam models. Generally, get the highest video resolution you can afford, because you'll probably use the camera for video chats. Consider the size of the camera, its attachment to your computer, and whether it has a microphone (you definitely need a microphone).

TIP

For instructions about opening apps such as the Camera app, visit Chapter 2.

2. The first time you use the Camera app, you may see a screen that says *Let Windows camera access your location?* Select Yes to continue.

3. The Camera app opens, and there's a good chance you recognize the face staring back at you, such as the example in **Figure 11-1.** To take a photo, smile and then click or tap the Take Photo button (it's on the right side of the screen). You may notice a blue box dancing around your face as you get ready to take the photo (see **Figure 11-1**). This is a focusing box that the camera uses, but it won't actually show up in your photo. You may hear a shutter click when you take the shot. Your photo is placed automatically in a folder called Camera Roll in the Pictures folder.

TIP

Select the Camera Roll icon in the lower-right corner of the Camera app to view the photo you most recently took. After that, you can select See All Photos to open the Photos App and view photos there. (Select the Back button in the upper-left corner of the screen to resume taking photos.)

4. To take a time-lapse photo, use the Photo Timer button on the left side of the screen. Select this button one, two, or three times to tell

the Camera app to wait 2, 5, or 10 seconds before taking the photo. Then click or tap the Camera button and wait 2, 5, or 10 seconds for the photo to be taken.

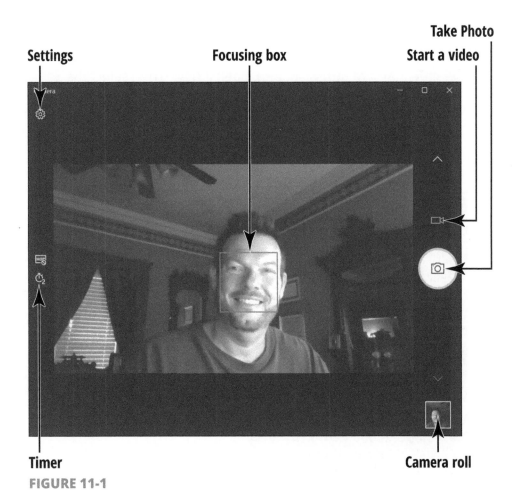

Settings · Focusing box · Take Photo · Start a video · Timer · Camera roll

FIGURE 11-1

5. Select the Video button (shown in **Figure 11-1**). Nothing happens until you click or tap the Video button a second time, at which point you're in moving pictures. A counter indicates the length of the video. Short is sweet in video. You can speak, too. Click or tap the Video button to stop the video. As with photos you take with the Camera app, videos are saved initially in the Camera Roll subfolder of the Pictures folder. Select the Camera button to turn off the video function and return to taking still photos (with the next click or tap).

Leave the Camera app open to explore the Camera settings, which we explain next.

Choose Camera Settings

1. If the Camera app isn't open, open it now. Then select the Settings button located in the upper-left corner of the Camera window (refer to **Figure 11-1**). The Settings panel opens, as shown in **Figure 11-2** and **Figure 11-3.** From here, you can tell the Camera app how to take photos and videos.

FIGURE 11-2

TIP

Depending on the size of your screen, you may need to scroll down to see all of the Settings panel. Just move your mouse over the panel, and a scroll bar appears on the right side.

Camera

← SETTINGS

Videos

Video quality

720p 16:9 30fps ⌄

Flicker reduction

Auto ⌄

Related settings

Choose whether camera can use location info

Change privacy settings

Try out new experimental features
Help test features in their preliminary stages. The feature(s)
may change or be removed at any time.

⦿⃝ Off

FIGURE 11-3

2. Turn Pro Mode on if you want to choose a brightness setting for your photos. After you turn on Pro Mode, a Brightness icon appears on the left side of the Camera app screen. Selecting this icon makes a slider appear so that you can adjust the brightness of photos from 100 percent (very bright) to 0 percent (very dark).

3. Choose a Framing Grid option to help aim the camera. For example, the Crosshairs option places crossing lines in the center of the screen so that you know where the center of your photos will be.

4. Under Photos, decide how you want to take photos:

- **Photo quality:** These options determine the aspect ratio and size of photos in pixels. For example, you can shoot narrow (the 4:3 options) or widescreen (the 16:9 options) photos.

- **Time lapse:** Choose this option when you want to take a series of time-lapse photos (the previous topic in this chapter explains how to take a time-lapse photo). For example, with this option selected and the Photo Timer set to 2 seconds, the Camera app takes a time-lapse photo every 2 seconds until you click or tap the Camera button a second time, which tells the app to stop taking photos.

5. Under Videos, decide how you want to take videos:

- **Video quality:** These options determine the aspect ratio and the number of pixels that appear in video frames. Notice that the fps (frames per second) setting in each option is the same, 30. *Aspect ratio* refers to the proportional relationship between a video's width and height. The most common aspect ratios are 16:9 (wide-screen) and 4:3 (old-fashioned TV). A *pixel* is one point of color. Suffice it to say, the more pixels in the setting, the clearer the video is.

- **Flicker reduction:** These options reduce the flickering that can occur when video is taken under fluorescent light. The 50 Hz (hertz) setting applies to the U.S. and Canada, where the AC (alternating electrical current) runs at 50 Hz; the 60 Hz setting applies to the rest of the world.

6. Return to the Camera app by clicking or tapping the Camera app screen.

7. Return to the Start screen. See the section "View Photos with the Photos App" to see and maybe edit the photos you just took.

Copy Photos from Your Camera

1. If your digital camera came with a cable, connect that cable to the camera, connect the other end of the cable to a USB port on your computer, and turn your camera on. If your laptop or tablet has a built-in card slot, you can take the memory card out of the camera and insert it in that slot.

2. If your computer doesn't have a built-in card slot, consider buying a small memory card reader that plugs into your computer and works with your camera's memory card. You don't need a multicard reader, just a reader with a single slot the size of your camera card. We consider a card reader to be more convenient than using a cable.

The techniques described here for importing photos from a camera also apply to videos. Follow these steps as well to copy videos from your digital camera to your computer.

3. Windows 11 detects your camera and may briefly display a notification indicating *Select to choose what happens with this device.* If you're quick enough to tap or click this notification, Windows 11 displays your choices, as shown in **Figure 11-4.** Select Open Device to View Files, if that option is available, to open File Explorer to the photos on your camera, as shown in **Figure 11-5.** Then skip to Step 4. If you didn't catch the notification in time or you didn't see the notification, continue to Step 4 to import photos another way.

TIP

See Chapter 3 for information on increasing the time a notification remains onscreen.

FIGURE 11-4

4. Select the File Explorer icon on the taskbar to start File Explorer. Then look for and select your camera in the navigation pane on the left side of the screen (refer to **Figure 11-5**). Your camera is located under *This PC* in the Navigation pane.

TIP

See Chapter 15 for detailed information about using File Explorer.

5. In File Explorer, select the photos you want to copy from your camera to your computer. Depending on your camera setup, the photos may be in a subfolder (refer to **Figure 11-5**). Use these techniques to select the photos:

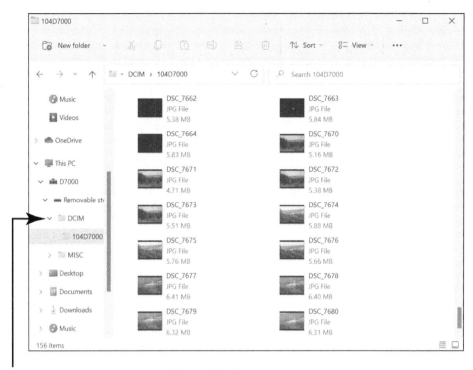

Select your camera or one of its subfolders

FIGURE 11-5

- Hold down the Ctrl button and click or tap photos to select them individually.

- Press Ctrl+A to select all the photos in a folder.

6. Copy the photos you selected so that you can paste them into a folder on your computer. Use one of these techniques to copy the photos:

- Press Ctrl+C.

- On the toolbar in File Explorer, select the Copy button.

7. In File Explorer, select the folder where you want to copy your photos.

TIP

If you want to be able to view and edit your photos with the Photos app, select the Pictures folder or one of its subfolders and copy the photos there.

TIP

Organizing photos can be a challenge. If you dump more than a few dozen photos into the Pictures folder without using folders, finding a specific photo later will be difficult. Using folders with unintelligible names doesn't help, either. Most of our folder names are based on the year and month (such as 2021–06) or the subject or location where we took the photos. You can name your folders whatever you want, but choose names that are meaningful to you and will help you find specific photos later.

8. Paste the photos into the folder you selected in Step 7. Use one of these techniques:

- Press Ctrl+V.

- On the toolbar in File Explorer, select the Paste button.

TIP

If you often import photos from a digital camera or memory card, you can speed the process by telling Windows to open File Explorer immediately when you attach your camera or memory card to your computer. Select the Search icon on the taskbar and type **autoplay settings**; and then select that option when it appears in your search. The AutoPlay window opens, as shown in **Figure 11-6.** In the Memory Card and Camera sections, choose Open Folder (or Open Device) to View Files (File Explorer).

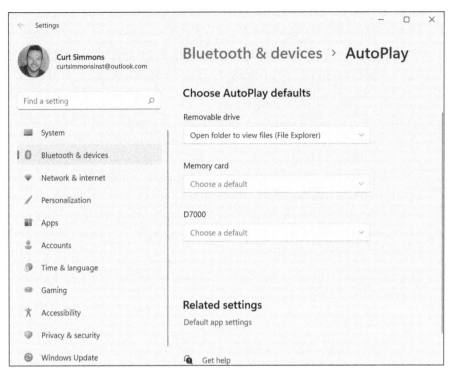

FIGURE 11-6

View Photos with the Photos App

1. To see photos on your computer, select the Search icon on the task-bar and type **photos** in the Search box; then select the Photos app in the search results. The Photos screen opens to the Collection window, as shown in **Figure 11-7.** It shows the contents of the Pictures folder (and its subfolders) on your computer. Your screen will look different.

2. If you don't see any pictures, go to one of these earlier sections:

- If you have a digital camera, see "Copy Photos from Your Camera."

- If your computer has a built-in camera or a webcam, see "Take Photos (and Videos) with Your Computer."

3. If you see pictures, either scroll down or drag the timeline on the right side of the screen to see more photos. Photos are presented by the dates on which they were taken.

View options

Timeline

FIGURE 11-7

TIP

You can take a screenshot — a picture of the current screen — by pressing ■+Print Screen. (A touchscreen or mouse equivalent is not available.) The screen dims slightly to indicate the capture, which is stored automatically in the Screenshots folder in your Pictures folder. Use this technique to create your own documentation of problems or something you want to see again later.

4. Select the View Large, View Medium, and View Small button. These buttons are located in the upper-right corner of the screen (refer to **Figure 11-7**). Use them to examine and locate photos.

5. Select any photo to display it full size in the Photos app window, as shown in **Figure 11-8.** When looking at a single photo, you can scroll from photo to photo by selecting the Previous or Next button or by clicking or tapping the right edge or left edge of the screen. You can also press the PageDown key (forward) or PageUp key (back). Note that the Previous and Next buttons don't appear until you hover or touch right or left sides of the window.

Return to photo collection   **Toolbar**   **See More button**

FIGURE 11-8

6. Select the See More button (the three dots on the right side of the toolbar; refer to **Figure 11-8**) and select Slideshow on the menu that appears. The photos appear in succession. Stop the slide show by clicking or tapping a photo.

7. Zoom in and out on a photo using one of these methods (repeat to zoom in or out more):

- **Mouse:** Click the Zoom In or Zoom Out buttons on the toolbar to zoom in or out on your photo.

- **Touchscreen:** Touch two fingers on the screen. Move your fingers apart to zoom in. Pinch your fingers closer together to zoom out.

- **Keyboard:** Press Ctrl+plus sign (actually, press the equal sign — no need to press the Shift key) to zoom in. Press Ctrl+minus sign to zoom out.

TIP

Zoom in to see part of a photo made larger.

Use the Delete button in the toolbar to delete a photo. (See Chapter 15 for information on undeleting files.)

8. Select the arrow on the left side of the toolbar to return to your photo collection. In this view, you can see more than one photo.

TIP The Photos app offers many options for editing photos. See "Edit Photos Using the Photos App," later in this chapter.

Organizing Photos Using the Photos App

1. In the Photos app, photos are organized by date in the Collection window (refer to **Figure 11-7**), but you can organize photos other ways as well, starting in the Albums, People, or Folders window. Organize photos so that you can find and view them easily. Select the Albums tab to open the Albums window. If you've already spent some time with the Photos app, the app may have created albums for you (each album comprises photos taken on the same day). If you asked for some photos to be "favorites" (you learn how to shortly), the Photos app creates a Favorites album so that you can view these photos. If you've never used the Photos app, you won't see anything in the Albums window except the option to create a new album.

2. Select New Album to create a new album for storing photos. The Create New Album window opens (see **Figure 11-9**). Scroll through and select photos you want for your album by selecting the check boxes on photos. Then select the Create button. The Album window opens, and you see a slide-show presentation of the photos you selected for the album. Select the Back button to return to the Albums window.

TIP To remove an album from the Albums window, click or tap to select it. Then select the Remove button. To change an album's cover photo, right-click or tap and hold the album, choose Change Your Photo on the pop-up menu, select a photo in the Choose a Cover Photo window, and select the Done button.

3. Select the People tab to open the People window, as shown in **Figure 11-10.** It may interest you to know that Windows 11 facial recognition software is capable of recognizing faces in photos. If you store a number of photos in the Photos app, the People window shows the faces of people that Windows 11 recognizes. By attaching names to the faces that are recognized, you can identify photo subjects by name and in so doing search for photos of people in the Photos app. To attach a name to photos that Windows 11 recognizes,

select Add Name below a face (refer to **Figure 11-10**). Then type a name or choose one from the Contacts list that appears, and press Enter.

Select photos for your album   Select Create

FIGURE 11-9

4. To find photos of someone who is recognized, click or tap in the Search box. A drop-down menu with names appears (refer to **Figure 11-10**). Select a name to see all the person's photos.

TIP

If you object to Windows 11 butting into your private life with its facial recognition software, select the See More button in the upper-right corner of the screen and choose Settings on the drop-down menu. Then, in the Settings window, scroll to the People setting and turn the People setting off.

5. Locate and select a photo you particularly like to display it full size. Then select the Add to Favorites button (the heart-shaped button along the top of the screen). Selecting Add to Favorites places a link to the photo in the Favorites album. Select the Back button, visit the Albums tab, and select the Favorites album. You see photos you deemed "favorites." Add a photo to the Favorites folder so that you can find it quickly.

To tag a photo, add a name   **Search for people you have tagged**

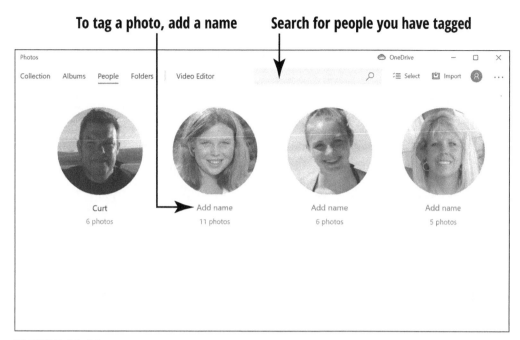

FIGURE 11-10

6. Select the Folders tab to open the Folders window. By default, the Photos app keeps track of photos found in the Pictures folder (and its subfolders) on your computer. Suppose you want to manage photos in other folders with the Photos app. Select Add a Folder. The Select Folder dialog box appears. Select a Folder and then select the Add Folder to Pictures button. Photos in the folder you select are now available for viewing in the Photos app.

Edit Photos Using the Photos App

1. You can use the Photos app to change a photo's size or appearance. In the Photos app, click or tap the photo you want to edit so that it fills the screen; then select the Edit & Create button, and choose Edit on the drop-down menu. The Editing window opens, as shown in **Figure 11-11.** This window offers many tools for touching up photos. To edit a photo, select a retouching method along the top of the screen (Crop & Rotate, Filters, or Adjustments) and then choose editing options.

Undo   Filter intensity

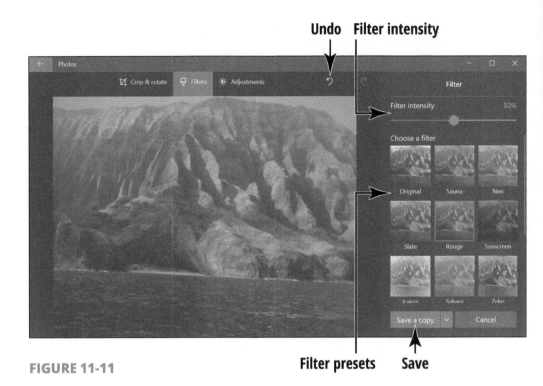

FIGURE 11-11

Filter presets   Save

2. Using the Filters method, you can take advantage of preset edits to photographs. After you select Filters, filter presets appear on the right side of the screen (refer to **Figure 11-11**). They offer the chance to change the light and color of a photograph in one fell swoop. Drag the Filter Intensity slider at the top of the screen to further change the look of a photograph.

TIP

If you make mincemeat of a photograph and you want the original version back, select the Undo button. Selecting this button restores a photograph to its original, pristine state.

3. Choose the Adjustments method to alter the brightness and color of your photograph (refer to **Figure 11-12**). By dragging sliders on the right side of the screen, you can do much to change the appearance of a photo.

4. Many photos can be improved by *cropping,* which involves cutting out distracting elements and keeping just part of the photo. You might crop a photo to concentrate on its most important part. To crop, select the Crop & Rotate method at the top of the screen. Then either select an aspect ratio or choose the Custom aspect ratio and drag

the size or corners of the picture box until the photo is to your liking. The box tells you which parts will remain after cropping, as shown in **Figure 11-13.**

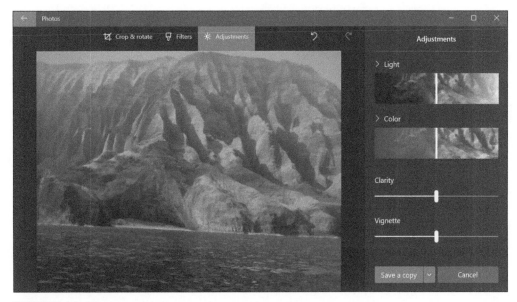

FIGURE 11-12

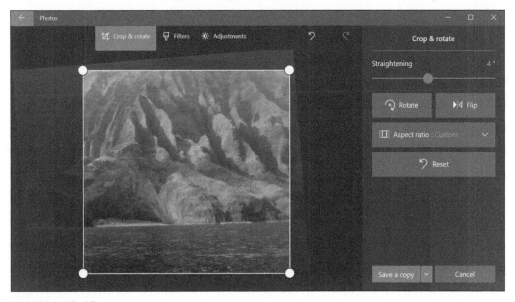

FIGURE 11-13

TIP

Sometimes when you import photos from a digital camera or scanner, they arrive askew because the photographer turned the camera the wrong way when shooting. You can correct this error by selecting Crop & Rotate and then selecting the Rotate button until the photo turns right-side-up.

5. Select Save a Copy when you finish doctoring your photograph to keep the original photograph as well as the edited version.

Print Photos Using the Photos App

1. If necessary, connect the cable from your printer to the USB port on your computer and make sure that your printer is turned on.

2. In the Photos app, select the photo you want to print using a click or tap to display the photo at full screen.

3. Select the Print button on the right side of the toolbar (it looks like a printer; refer to **Figure 11-8**). The Print window appears, as shown in **Figure 11-14.**

TIP

Here's a handy keyboard shortcut that works in any app: Press Ctrl+P to open the Print window straightaway.

4. Select a printer. Note the preview. Select the Print button.

TIP

Chapter 14 explains printing in detail.

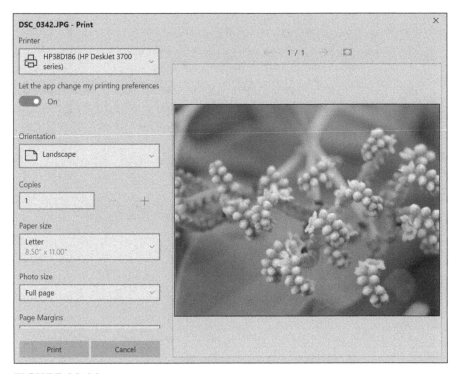

FIGURE 11-14

Scan Photos and Documents

1. Scan photos and documents to preserve them in digital form. Old family photos and heirloom documents such as birth certificates are candidates for scanning. *Scanning* means to render a paper photo-graph or document as a computer file so that you can post it on the Internet or send it by email. To scan photos and documents, you must have a scanner or a printer that is capable of scanning (most printers have a scanning capability). To start, connect the cable from your scanner or printer to the USB port on your computer. Make sure that your scanner or printer is turned on. If you're using a printer, put it in Scan mode (you'll need to check your printer documentation for instructions and details).

2. Put the item you want to scan in your scanner or printer.

3. Select the Search icon on the taskbar and type **scan** in the Search box. In the search results, select Windows Fax and Scan. The Windows Fax and Scan app opens.

4. In the navigation window, select the New Scan button on the toolbar. The New Scan dialog box appears (see **Figure 11-15**). Starting here, you can tell Windows whether you want a high-quality or low-quality scan. On the Color Format menu, for example, choose Grayscale or Black and White instead of Color if you're scanning a black-and-white document. Choose a Resolution (DPI) higher than 300 to render your photo or document in greater detail.

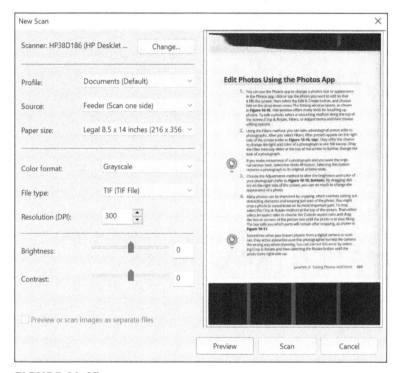

FIGURE 11-15

TIP

Unless you're scanning more than one item, deselect the Preview or Scan Images as Separate Files option. This option creates a separate digital file for each item in your scanner.

5. Select the Preview button. The preview window shows what your photo or document will look like when it is scanned. If you don't like what you see, play with the settings in the New Scan dialog and select the Preview button again.

6. Select the Scan button when you're ready to scan your document. Your scanned photo or document appears in the Windows Fax and Scan window with the generic filename "Image."

7. Select your newly scanned image and then select the Save As button. The Save As dialog box appears. Enter a descriptive name for the item, choose a folder to store it in, and select the Save button.

Chapter 15 explains how to save files.

TIP

Chapter 12

Enjoying Music and Videos

The term *media* on computers refers to something other than text. Audio and video are examples of media. *Audio* is a catchall term for music and other sound files, such as books on CD. *Video* includes files that you can shoot with your digital camera as well as Hollywood blockbusters.

The delivery of music has come a long way from Edison's wax cylinder or even vinyl LPs. Nowadays, music is almost entirely digital (although vinyl has been making a comeback in some circles). The Groove Music app lets you play your music collection and makes it easy to explore new music. Use Windows Media Player to play audio CDs, to copy audio files to your computer, and to burn songs onto a CD.

The Movies & TV app is similar to the Groove Music app, but it's for video instead of music. You can use it to play your homemade videos.

In this chapter, you play a music CD, copy CDs to your computer for easier access, and burn a CD. You also explore the Groove Music and Movies & TV apps. Finally, you discover how to make a voice recording.

TIP You'll need a different app to play DVDs. Search Microsoft Store for Media Center or for a DVD player app. See Chapter 9 for more information on Microsoft Store.

TIP Popular alternatives for music or videos include Hulu, iTunes, Netflix, Amazon, and YouTube. All of these are accessible through the web browser. Search Microsoft Store for related apps.

Play and Copy Music from a CD

1. If you have a CD or DVD disc drive or slot on your computer, insert a music CD, label side up for horizontal drives. (Vertical drives are less predictable.)

2. Windows 11 detects your CD and may briefly display this notification: *Choose what to do with audio CDs.* Windows 11 displays your choices, as shown in **Figure 12-1.** Select Windows Media Player and then skip to Step 5.

FIGURE 12-1

3. If you see the Welcome to Windows Media Player screen, select Recommended Settings and then select Finish. The Windows Media Player plays your music.

4. For access to more options, select Switch to Library, the small button under the *X* in Windows Media Player, as shown in **Figure 12-2.**

5. Note the following controls at the bottom of the Windows Media Player, as shown in **Figure 12-3:**

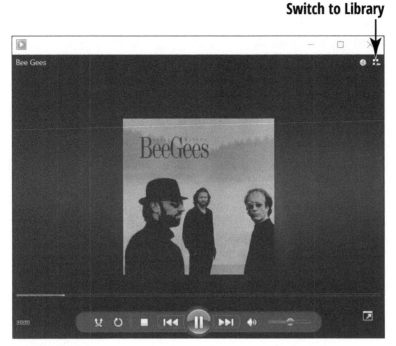

Switch to Library

FIGURE 12-2

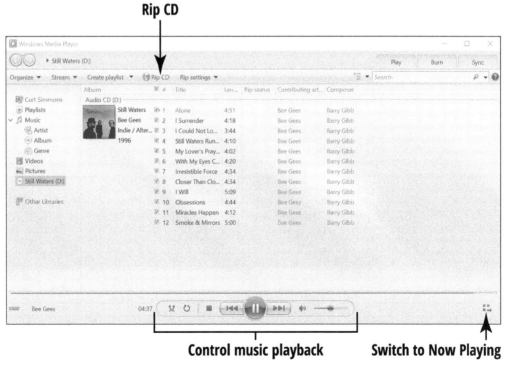

Rip CD

Control music playback   **Switch to Now Playing**

FIGURE 12-3

- **Shuffle:** Select this button to turn on *shuffle,* which randomly mixes the tracks you play. Select again to turn off shuffle, and the tracks play in the order in which they appear onscreen.

- **Repeat:** Select this button to play all the tracks again after all have played. Select again to turn off the repeat function.

- **Stop:** Select to stop playing.

- **Previous:** Select this button to skip to the previous track. Select and hold to rewind to an earlier point in the track.

- **Play/Pause:** Select the button with two vertical lines to pause play mid-track. Select the same button (now with a triangle pointing to the right) to resume playing from the point you paused.

- **Next:** Select this button to skip to the next track. Select and hold to fast-forward through the track.

- **Mute/Unmute:** Select this button to silence the player. Although the track continues to play, you won't hear it. When Mute is on, a red circle with a slash appears next to the speaker icon. Select the button again to hear the track.

- **Volume:** Drag the slider to the left to decrease or to the right to increase the volume of the track. Your speakers may also have a manual volume control. Windows 11 has a separate volume control in the taskbar, as well.

- **Switch to Now Playing:** You select this button, which is located in the lower-right corner of the window, to reduce the player to a small size (refer to **Figure 12-2**).

6. To copy the CD tracks to your Music library, select Rip CD. This button is located on the toolbar along the top of the screen, as shown in **Figure 12-3.** Ripping a song on a CD takes about a minute. As each song is ripped, the words *Ripped to Library* appear in the Rip Status column of the Windows Media Player.

TIP

If you plan to rip a lot of CDs, select Rip Settings ⇨ Rip CD Automatically and Rip Settings ⇨ Eject CD After Ripping. Just inserting the CD will copy files to your Music library as it plays the CD. Audiophiles should choose Rip Settings ⇨ Audio Quality ⇨ 320 Kbps (Best Quality). As well, audiophiles may consider choosing Format ⇨ WAV, ALAC, or FLAC. These lossless formats are superior to the MP3 format, although fewer devices play files in lossless formats.

7. When the copying process finishes, remove your CD. To play this music in the future, start Windows Media Player, choose Artist, Album, or Genre under Music, select the CD you ripped, select a song title, and select the Play button.

 Songs you rip from a CD are kept on your computer in a subfolder of the Music folder. To copy or move music that you ripped from a CD, open the Music folder, open the subfolder named after the CD you ripped, and copy or move the files.

TIP

 You can play anything in the Music library by using the Groove Music app. See the next section, "Listen to the Groove Music App."

TIP

 Pin the Windows Media Player to the taskbar for easy access: Click the right mouse button over the icon in the taskbar, or tap and hold until a box appears, and then release. On the menu that pops up, select Pin to Taskbar. You can also search for Windows Media Player using the Search icon on the taskbar. See Chapter 2 for information on pinning apps.

TIP

 If your music CD doesn't play automatically the next time you insert one, you can tell Windows 11 how to handle audio CDs. On the taskbar, select the Search icon and then type **autoplay** in the Search box. In the search results, select AutoPlay Settings. The Settings screen opens to the AutoPlay settings. Turn on the Use AutoPlay for All Media and Devices option to make CDs play automatically.

TIP

Listen to the Groove Music App

1. Select the Search icon on the taskbar and type **groove** in the Search box. Select Groove Music in the search results. The Groove Music app home screen appears, as shown in **Figure 12-4.** The Groove Music app enables you to play songs that you ripped (copied) from CDs,

songs that you purchased from iTunes and other music purveyors, and playlists that you created from songs you own.

TIP

If you see *Can't sign in,* you're not connected through a Microsoft Account. You'll still be able to do the steps in this section, though. You need to sign in only to buy music.

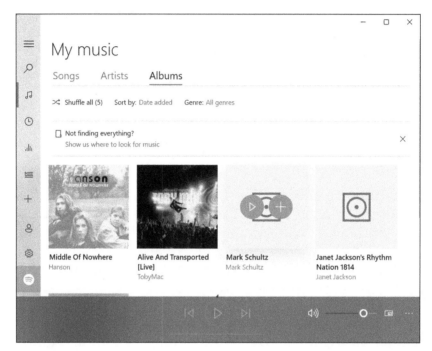

FIGURE 12-4

2. Initially, the Groove Music app recognizes music files kept in these folders on your computer: C:\Users*Your Name*\OneDrive\Music and C:\Users*Your Name*\Music. Maybe you keep your music files in other folders. To tell the Groove Music app where your music files are, tap or click the Settings button (see **Figure 12-5**). In the Settings screen, select the *Choose Where We Look for Music* link. The Build Your Collection dialog box appears, as shown in **Figure 12-6.** Select the Add Folder button, choose a folder in the Select Folder dialog box, and tap or click the Add This Folder to Music button. Then select Done. Repeat these instructions to add all the folders where you store music files.

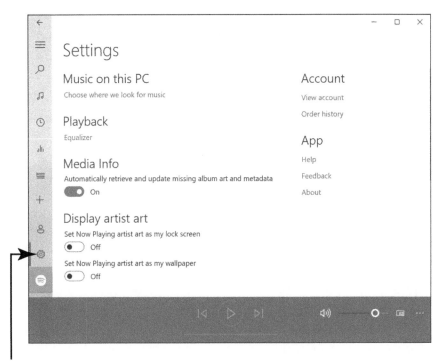

Settings

FIGURE 12-5

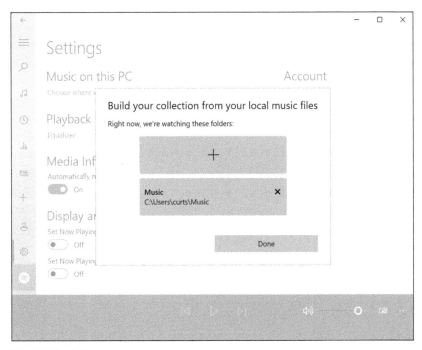

FIGURE 12-6

CHAPTER 12 **Enjoying Music and Videos** 233

3. To find the song you want to hear, choose an option in the navigation pane and get to it:

- **Search:** Type the name of an artist, album, or song in the Search box. As you type, suggested matches appear below the Search box. For now, ignore these suggestions (but take advantage of them in the future to save typing). Instead, select the magnifying glass or press Enter. Search results appear.

- **My Music:** After you make this choice, select Songs, Artists, or Albums (refer to **Figure 12-4**) to search from a song list, a list of artists, or a list of albums.

- **Recent Plays:** A list of albums you played recently appears. Scroll to find an album.

TIP

If the search results list is a long one, you can narrow your search with the Sort By and Genre options (refer to **Figure 12-4**). Sorting opens a pop-up list whereby you can arrange the results in alphabetical order, date-added order, or by artist or album. Choosing Genre opens a pop-up list on which you can choose a particular genre of music.

4. Select a song or album in the search results. Note the following options (not all of which may appear on the current screen):

- **Play:** Plays the song. Notice the controls at the bottom of the screen for pausing the song and controlling the volume (refer to **Figure 12-7**).

- **Play All:** Plays the songs on the album, starting with song 1.

- **Add To:** Opens a pop-up menu with playlists, and you can add the song to one of those playlists (see "Create and Manage Playlists," the next topic in this chapter).

TIP

You can listen to music as you conduct a search. To return to the song or album that is currently playing, select the Now Playing button.

Play All   **Add To**

Playback controls

FIGURE 12-7

5. To control the volume level on your speakers, use either the volume controls in Groove Music or the Windows 11 volume controls. Select the Speakers icon in the lower-right corner of the Groove Music screen or the Windows 11 screen. A volume slider appears. Slide the control to adjust the volume. Select the speaker icon on the right side of the slider to mute or unmute all sounds.

6. Switch back to the Start screen. The Groove Music app continues to play. The Groove Music icon on the Start screen displays the album art and title of the current song. Select the Groove Music tile to return to the app.

7. Leave the Groove Music app open if you care to discover how to create a playlist, the next topic in this chapter.

Create and Manage Playlists

1. Create a playlist with the Groove Music app to play your favorite songs — songs from different artists — one after the other. After you create the list, you select it in the Groove Music app to play it. To create a playlist in the Groove Music app, select the Create New Playlist button (a plus sign) in the left navigation pane. You see the Name This Playlist dialog box, shown in **Figure 12-8.** Enter a descriptive name for your playlist, and select Create Playlist.

FIGURE 12-8

2. The Groove Music App has buttons to add songs to a playlist. Next time you're listening to a song or album and you think, "I like that song; it should be on a playlist," do one of the following to add the song to a playlist, as shown in **Figure 12-9:**

   - **Add a song:** Select a song's Add To button and select a playlist on the pop-up menu that appears.

   - **Add all the songs on an album:** Select the Add To button on the album name and select a playlist on the pop-up menu.

FIGURE 12-9

See "Listen to the Groove Music App," earlier in this chapter, if you need help finding and listening to music on your computer.

3. To play the songs on a playlist, select the Playlist button on the left side of the screen; then choose the playlist you want to play. The playlist opens, as shown in **Figure 12-10.** Select the Play All button to play the songs in order from first to last. Of course, you can select any song on the list and select its Play button to play it.

4. To change the order of songs on a playlist or remove a song, follow these instructions:

   - **Change the order of songs:** Drag a song higher or lower on the list with your mouse or finger.

   - **Remove a song:** Select the song in the playlist and then select the Delete from Playlist button.

To delete a playlist, open it. Then select the Delete button.

TIP

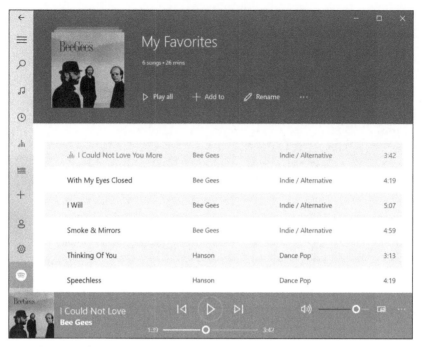

FIGURE 12-10

Watch Videos

1. Select the Search icon on the taskbar and type **movies** in the Search box. Select the Movies & TV app that appears. Microsoft designed this app for watching movies from the Microsoft Store and for watching homemade videos. To watch homemade videos, select the Personal tab on the menu bar. The Videos screen appears, as shown in **Figure 12-11.** Use the Video screen to watch videos.

2. Unless you store your videos in the `C:\Users\`*Your Name*`\Videos` folder, you don't see any videos. To tell the Video app where you store your videos, select the Add Folders button. You come to the Build Your Collection screen. Select the Add button (the plus sign), and in the Select Folder dialog box, select the folder where videos are stored and then select the Add This Folder to Video button. Repeat these instructions to add all the folders where you store videos.

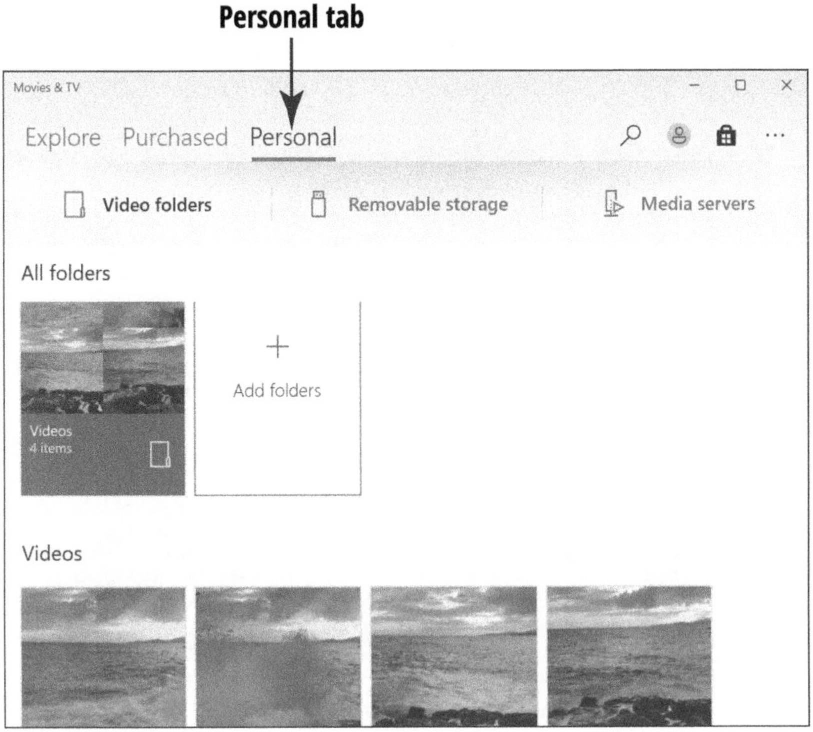

Personal tab

FIGURE 12-11

3. To play a video, select it on the home screen. The video appears in the video player screen, as shown in **Figure 12-12.** Along the bottom of this screen are controls for pausing, playing, and changing the volume level. Drag the slider below the video screen to rewind and fast-forward.

4. Select the Back button (located in the upper-left corner of the screen) to return to the Movies & TV home screen.

Drag to rewind or fast-forward   **Video controls**

FIGURE 12-12

Burn a CD

1. *Burning a CD* means to copy songs from a computer to a CD. Burn a CD so that you can play the songs on your car's CD player or pass along your favorite songs to a friend. To burn a CD, start by opening Windows Media Player, as shown in **Figure 12-13.**

TIP

Earlier in this chapter, "Play and Copy Music from a CD" explains Windows Media Player in detail.

TIP

To burn a CD, your computer needs an *optical disk drive,* a hardware device for playing and recording to CDs and DVDs. If your computer doesn't have one and you want to burn CDs, you can buy optical disk drives that plug into your computer.

Drag song titles over the playlist name

Create a playlist   **Change views**

FIGURE 12-13

2. Create a playlist so that Windows Media Player knows which songs to burn to the CD. To create a playlist, select the Create Playlist button and enter a name for the playlist where the words *Untitled Playlist* appear. The name you enter then appears under Playlists on the left side of the screen.

3. Locate songs for your playlist and drag and drop their titles one by one onto your playlist name. The words *Add to [Playlist Name]* appear when you drag a song title onto your list. Use these techniques to locate songs so that you can drag them in the playlist:

- **Browse for songs:** Under Music, select Artist, Album, or Genre on the left side of the screen (refer to **Figure 12-13**). A list of artists, albums, or genres appears.

- **Change views:** Open the Change Views drop-down menu and choose Icon, Title, or Details to make finding a song title easier.

- **Search for songs:** Enter a search term in the Search box.

4. Select your playlist on the left side of the screen. The songs you selected for the list appear, as shown in **Figure 12-13.** To listen to the songs on a playlist, select its name and click or tap the Play button at the bottom of the screen.

TIP

You can edit a playlist. While the list is playing, place more songs on the list (follow Step 3) or remove songs from the list by right-clicking and choosing Remove from List.

5. After you create a playlist, you can burn it to a CD. With Windows Media Player open, insert a blank CD in your computer's CD or DVD drive or an optical drive plugged into your computer. Then select the Burn tab (refer to **Figure 12-14**) and drag a playlist from the left side of the screen to Burn List on the right.

Start Burn   **Burn List**

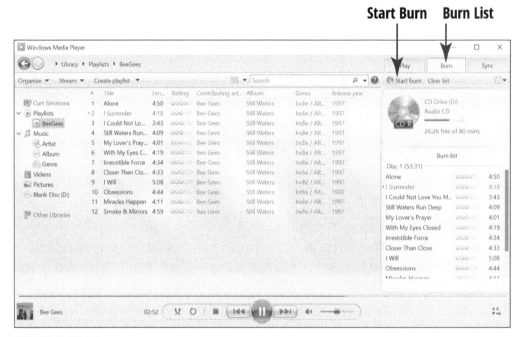

FIGURE 12-14

TIP

You can change the order of the songs before you burn them onto a CD. Drag song titles up or down the list to change the song order.

6. Note the Audio CD bar on the right side of the screen. It tells you whether there is enough disc space to burn all the songs on the list.

If there is enough space, select the Start Burn button. It can take a few minutes to burn a CD. When the job is complete, your computer ejects the CD. We suggest playing it to see whether the songs copied correctly to the CD. While you're at it, label the CD by writing on it with a felt-tip (not a ballpoint) pen. Label the top of the CD, not the bottom.

Make a Voice Recording

1. Open the Voice Recorder app to record your voice. As shown in **Figure 12-15,** Voice Recorder lists recordings you already made (if you made any recordings).

TIP

It goes without saying, but your computer needs a built-in microphone or microphone jack for plugging in a microphone if you want to make voice recordings.

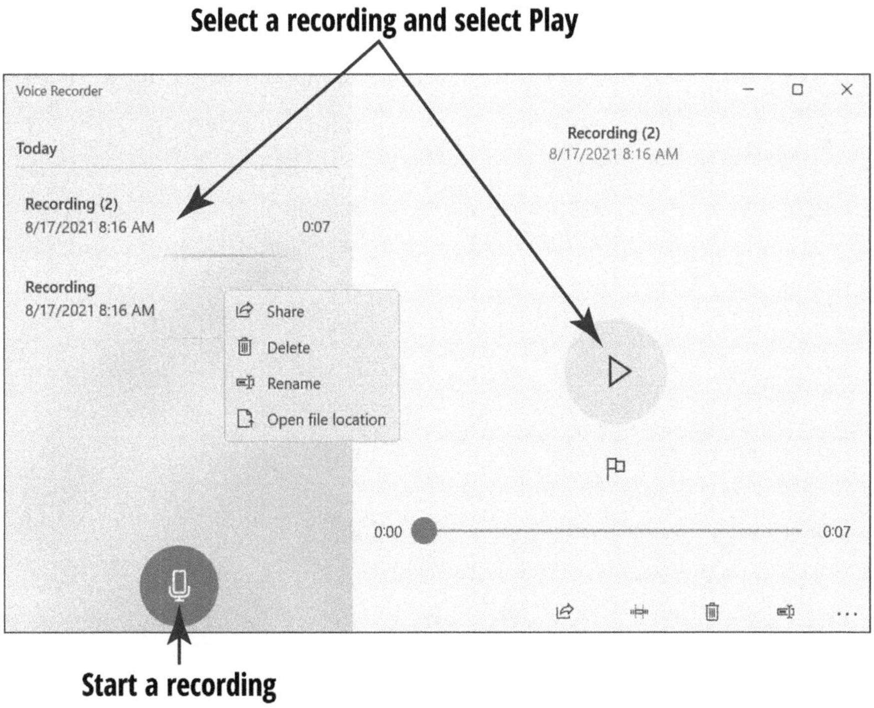

Select a recording and select Play

Voice Recorder

Today

Recording (2)
8/17/2021 8:16 AM   0:07

Recording
8/17/2021 8:16 AM

     ↪ Share

     🗑 Delete

     ⊟ Rename

     ⬀ Open file location

Recording (2)
8/17/2021 8:16 AM

▷

⚑

0:00 ●━━━━━━━━━━━━━━━ 0:07

Start a recording

FIGURE 12-15

TIP

The first time you run Voice Recorder, a dialog box asks, "Let Voice Recorder access your documents library?" Select Yes. To change this later, go to the Settings app.

2. Select the Record button (or press Ctrl+R) to begin a recording. A timer tells you how long your recording is. You can select the Pause/Resume Recording button to temporarily stop a recording. Select the Pause/Resume Recording button again to continue recording.

3. Select the Stop Recording button (located in the middle of the screen) to end a recording. The name of your recording appears in the list under the generic name *Recording*. To give it a more descriptive name, right-click or tap to display the recording's context menu (refer to **Figure 12-15**), choose Rename, enter a name, and select the Rename button.

To delete a recording, display its context menu and choose Delete.

TIP

TIP

To access a Sound Recorder file, perhaps to copy or move it, display its context menu (refer to **Figure 12-15**) and choose Open File Location. The Sound Recordings folder opens. From here, you can copy or move the file. See Chapter 15 if you need instructions for handling files.

4

Beyond the Basics

Fine-tune Windows 11.

Manage printers and other peripherals.

Manipulate files and folders.

Keep backup copies of files and folders.

Chapter **13**

Maintaining Windows 11

Windows 11 is a bit like a car. To make it run well, you have to maintain it.

Windows 11 uses several tools to keep you informed of issues that pertain to your computer's health. The Reliability Monitor can help you pinpoint problems with hardware and software.

Machines such as can openers aren't getting any smarter. Your computer, however, can be programmed to do something it's never done before. To make your computer capable of doing new things, you install new programs. On the other hand, your computer may have some programs that you'll never use and wouldn't miss. You don't have to get rid of them, but doing so is easy enough and can make your computer run better.

In this chapter, you work with the Windows Security screen to check your computer's health status. You also install a program and, optionally, uninstall one. This chapter also shows how to make your computer work faster by controlling the startup apps, defragmenting your hard drive, and deleting unnecessary system files. Finally, you discover how to protect your computer against spyware, viruses, and other foreign invaders.

Install a New Program

1. You can install software that you download from the web or from a CD or DVD. (Windows 11 apps must be installed through the Microsoft Store.) In this set of steps, you install Audacity, a free program for editing audio files. To install Audacity, open your browser and enter **audacityteam.org/download** in the address bar. You land on the web page for downloading Audacity. (See Chapter 6 for information on using the Edge browser.)

TIP

If a website offers to install a program automatically, regard that suggestion with suspicion. It may be legitimate or it may be malevolent. Decline downloads from sources that you don't know and trust already.

2. Select the Audacity Installer link for Windows to go to the web page for downloading programs.

3. Select the Audacity Installer link to start the download; however, note that Edge may do this for you automatically. Either way, a Downloads pop-up menu appears, showing that the software has been downloaded. Choose Open File on this pop-up, shown in **Figure 13-1.**

4. The User Account Control dialog box appears. Select the Yes button to tell Windows you want to download and install the program. In the Select Setup Language dialog box, choose English and select OK.

TIP

Generally, if the installer for a program offers Express or Custom installation options, choose the Express option to let the installer set up the program without further input from you. The Custom or Advanced Settings option allows you to specify where to install the program and, perhaps, which parts of the program to install.

5. In the Setup – Audacity dialog boxes that appear, click Next until you come to the dialog box that asks whether you want to install the program. Select the Install button. In the Information dialog box shown in **Figure 13-2, top,** select the Next button. Last but not least, select the Finish button in the dialog box that tells you that Audacity has finished installing, as shown in **Figure 12-2, bottom.** The Audacity program opens on your screen.

Downloaded file

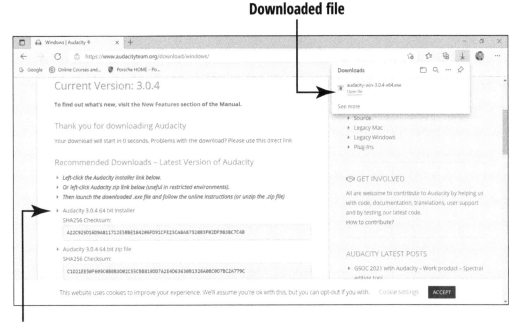

Select to download

FIGURE 13-1

TIP

Many programs try to connect to the Internet for updates during installation or when you run the installed program. The first time you run a program, you may be asked whether you want to register the program or configure some aspect of the program. Go with the default (assumed) responses, if you're not sure.

6. Close Audacity. You can do that by selecting the Close button (the X) in the upper-right corner of the screen.

TIP

In the unlikely event that you need to install a program that comes on a CD or DVD, insert the program disc into your computer's disc drive or tray, label side up (or, if your computer has a vertical disc slot, insert the disc with the label side facing left). The AutoPlay dialog box appears. Select the option to run Install or Setup. User Account Control may ask whether you really want to run this program. (Windows 11 tries to keep you from install-ing software unintentionally by asking for confirmation.)

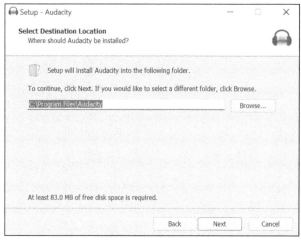

FIGURE 13-2

Remove a Programs

1. Unlike Windows 11 apps from the Microsoft Store, you install other programs from a website or even a CD. To see which programs are installed, open the Control Panel. Select the Search icon on the taskbar and type **control panel**; then select Control Panel in the Search results.

TIP

Many of the functions covered in this chapter are part of the Control Panel, which presents many functions for tweaking your computer setup.

2. In the Control Panel window (shown in **Figure 13-3**), under Programs, select Uninstall a Program. (You don't have to uninstall anything right now; you can simply see what the option offers.) You see the Programs and Features window.

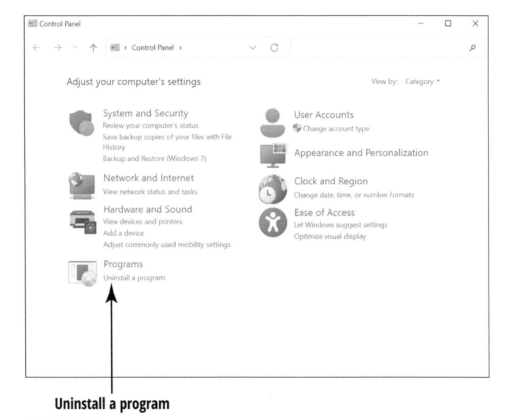

Uninstall a program

FIGURE 13-3

3. The Programs and Features window lists programs installed on Windows 11, but it does not list your Windows 11 apps. Initially, these programs are sorted by name. You may want to see the date you last used each program, because a program you haven't used in ages

may be a candidate for removal. (Otherwise, skip to Step 5.) Use one of the following techniques to display the date last used:

- **Mouse:** Move the mouse pointer over any column heading, such as Name. Click the right mouse button.

- **Touchscreen:** Tap and hold down on any column heading, such as Name. When a box appears around your fingertip, lift your finger.

4. On the context menu that appears, select More. In the Choose Details window (shown in **Figure 13-4**), select Last Used On and then select the OK button. The Last Used On column appears to the right of all the other columns.

Before you uninstall a program that you may want to reinstall later, make sure you have the installer file (or the installer CD or DVD for older programs) (or that you know where to download it from the web again). You have no undo option when you uninstall a program.

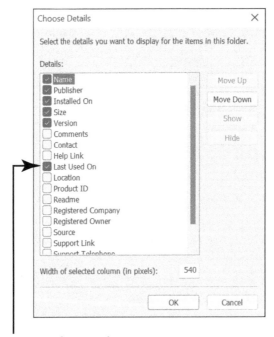

Last Used On option

FIGURE 13-4

5. To uninstall a program, select it. For this example, select Audacity, as shown in **Figure 13-5.** You don't have to uninstall Audacity, but if you do, you can follow the steps in the preceding section to reinstall it.

Uninstall option

FIGURE 13-5

6. Select the Uninstall button (located above the list of program names). Some programs offer the Change and Repair options as well as Uninstall. Repair or Change may be useful for a program that you want to keep but isn't running as expected. When Windows 11 asks whether you really to uninstall, select Yes or No.

7. If you uninstalled Audacity or another program, it will no longer appear in the Programs and Features window.

TIP

Just because you can uninstall a program doesn't mean you should. You can simply ignore programs you don't use. Look at a program's name, publisher, and date installed to determine whether you actually use a program. You may recognize a program you installed recently, as opposed to one installed before you got your computer. If you find that it's more productive to remove large programs than small ones, repeat Steps 3 and 4 to group by size.

After you install a program, take note of whether your computer runs differently. If your computer slows or behaves sluggishly, the program you recently installed could be the culprit. Uninstall it, if necessary.

Control Startup Apps

1. Startup apps are applications that start running when you turn on your computer. You might be surprised by how many startup apps run in the background without your knowing it. Occasionally you install new software or download a program from the Internet and discover that it runs automatically whether you like it or not. Sometimes the number of startup apps slows the computer down. To see which apps start running automatically when you turn on your computer, start the Task Manager with one of these techniques:

    -   **Keyboard:** Press Ctrl+Alt+Del and select Task Manager on the blue screen.

    -   **Search:** Select the search icon on the taskbar, enter **task** in the Search box, and then select the Task Manager in the search results.

2. Select More Details (if necessary) in the Task Manager window and then select the Startup tab, as shown in **Figure 13-6.** This is the list of applications that start when your computer starts.

3. Use one of these techniques to open your browser, go on the Internet, and find out what one of these startup applications does:

    -   **Mouse:** Right-click an application and choose Search Online.

    -   **Touchscreen:** Tap and hold down the name of an application. When a box appears around your fingertip, select Search Online.

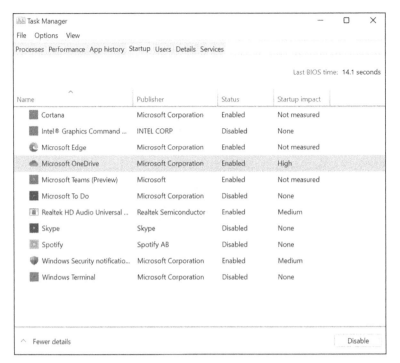

FIGURE 13-6

4. Notice the Startup Impact column in the Task Manager. Applications with a high impact such as Microsoft OneDrive may be slowing down your computer. If you decide that one of these applications is unwanted or is slowing you down, disable it. (Don't worry, you can enable it later on very easily, as we explain shortly.) To disable an application, select its name in the list and then select the Disable button.

TIP

To enable a startup application in the Task Manager (refer to **Figure 13-6**), select its name and then select the Enable button. The Status column in the Task Manager tells you whether a startup application has been disabled.

Explore System Information

1. Select the search icon on the taskbar and type **control panel** in the Search box.

2. Select Control Panel in the Search results. In the Control Panel, select System and Security.

3. In the System and Security window, select System. The System window shown in **Figure 13-7** appears. This screen is chock-full of information and functions. (You need to scroll down the screen to see everything.) Note each of the following areas onscreen:

   - **Windows edition:** Of course, you have Windows 11. However, Windows 11 comes in two primary editions: the Home edition, which you are likely to have, and the Pro edition, for computer professionals. You may find information here about so-called Service or Experience Packs, which are large collections of updates to Windows 11.

   - **Device Specifications:** This section displays details about your hardware, including the processor, the amount of installed memory (RAM), and other details.

   - **Computer name, domain, and workgroup settings:** This information pertains to your network, if you have one. If a computer can't connect to a network, the problem is often related to the name of the Workgroup (a network). The Change Settings function lets you change the Workgroup name to match other computers on the same network.

   TIP

   Windows 11 gives your computer a nondescript, generic name to begin with. To give your computer a more descriptive name, select Rename this PC at the very top of the System window (refer to Figure 13-7). Enter a new name in the Rename Your PC dialog box.

   - **Product key and activation:** In an effort to control software piracy involving bootlegged copies of Windows 11, each copy of Windows 11 must be activated. Odds are that you activated your copy the first time you started your computer.

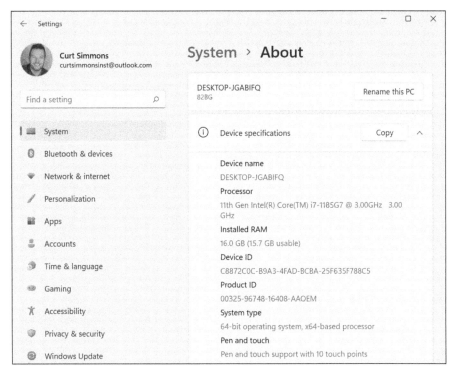

FIGURE 13-7

TIP

Don't be alarmed by the System information and options. If all goes well, you don't have to use most of what you find here. Some familiarity with this screen will be useful, however, if all doesn't go well later.

Check Your Security and Maintenance Status

1. In Control Panel, select System and Security, and on the next screen, choose Security and Maintenance (see **Figure 13-8**). The Security and Maintenance window appears, as shown in **Figure 13-9.**

TIP

Another way to open the Security and Maintenance window is to type **Security and Maintenance** using Search on the Windows taskbar.

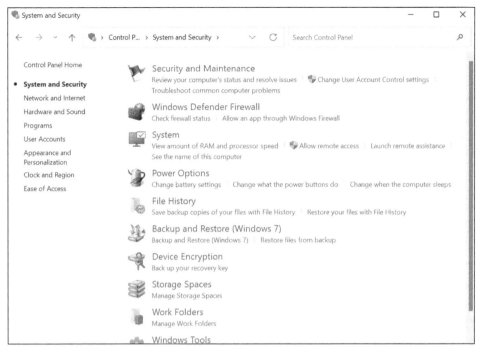

FIGURE 13-8

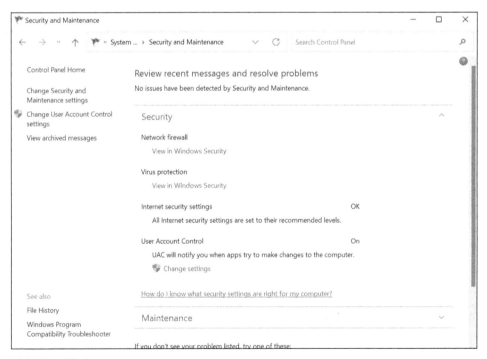

FIGURE 13-9

2. Note any message displayed under Review Recent Messages and Resolve Problems. Ideally, you see *No issues have been detected by Security and Maintenance.* If you see a message concerning a specific problem, select that message for more information.

3. Select the Security heading in the Security and Maintenance window. That section expands to detail security functions. Every option should display *On* or *OK*. Here's a brief description of each item under Security:

   - **Network Firewall:** The firewall scans Internet traffic and blocks activity from programs that don't have explicit permission to use Internet access. When you install a program that uses the Internet, you may be asked to approve the connection the first time. The safest practice is to reject online connections that you don't initiate or recognize.

   - **Virus Protection:** Having virus protection for your computer is essential. Windows Defender, an antivirus protection program, comes with Windows 11, although you can install another antivirus program if you want. Later in this chapter, "Protect Your Computer from Viruses and Other Threats" explains how the Windows Defender works.

   - **Internet Security Settings:** These settings pertain to your browser. The default settings may be adequate.

   - **User Account Control:** This function notifies you of programs that try to make changes to your system and requires that you confirm any such changes. In particular, UAC lets you know when a program tries to run or install software that may be malicious. When in doubt, say No or Cancel to UAC messages.

4. Select the Maintenance heading to see what that section includes. Functions under Maintenance consist of the following:

   - **Report Problems:** This setting is on, allowing Windows 11 to regularly check for solutions to problems it uncovers. (In Step 5, you run the Reliability History report from this part of the screen.)

   - **Automatic Maintenance:** Your computer automatically performs critical updates, security scans, and diagnostics each day.

If your computer is in a guest room or bedroom, you may want to change the Automatic Maintenance setting to run maintenance tasks at some time other than the default 3:00 a.m. Your computer may actually wake up at that hour for maintenance (although, if your computer is connected to a power strip, you can turn off the power strip and prevent your computer from turning on automatically in the middle of the night). If the computer can't run maintenance at the appointed hour, it will do so at the next opportunity.

- **File History:** See Chapter 16 for information on using the File History option, which is off by default.

- **Drive status:** *Drives* are hard disks inside or attached to your computer. Your documents, photos, and Windows 11 itself are stored on one or more drives. Ideally, the drive status is *All drives are working properly*. See Chapter 16 for information on backing up and restoring files.

- **Device software:** If a device on your computer needs a driver or other type of software to run properly, you are alerted here (if no drivers are needed, you see the words *No action needed*). Select Install Device Software to install the software.

The Security and Maintenance window is a troubleshooting tool, so you should check it if you have problems running Windows 11.

5. Under Report Problems, select View Reliability History. As shown in **Figure 13-10,** the Reliability Monitor screen graphs your computer's stability and indicates hardware and software problems, including those you may not be aware of. On this screen, red circles indicate critical events such as computer crashes; blue circles are information about software installation and updates; and yellow triangles indicate warnings about noncritical events (something that didn't crash the computer). Select a day in the graph to display details in the lower portion of the screen.

Reviewing the Reliability Monitor screen helps you distinguish between a one-time glitch and a recurring or worsening problem.

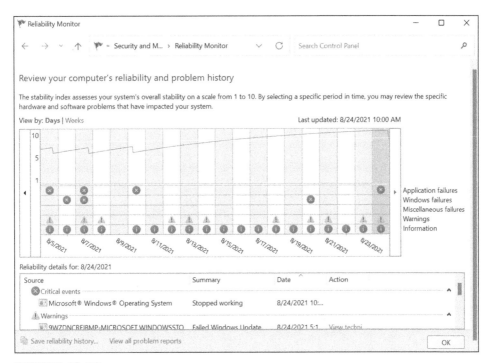

FIGURE 13-10

Speed Up Your Computer

1. Frequently used computers tend to slow down over time. However, Windows 11 offers a couple of administrative tools that can make your computer work faster. To find out what these tools are, select the Search icon on the taskbar and type **windows tools** in the Search box. Then select the Windows Tools app in the search results. The Windows Tools app shows you a folder of tools you can use, as shown in **Figure 13-11.**

2. Select Defragment and Optimize Drives. The Optimize Drives screen opens, as shown in **Figure 13-12.** When you save a file, the new data you recently added gets placed on the hard drive wherever Windows 11 can find room for it. Consequently, a file is stored in many different places. If you've used your computer for a long time, files become fragmented — the bits and pieces are spread all over the drive, and your computer has to work hard to assemble all the pieces to open a file.

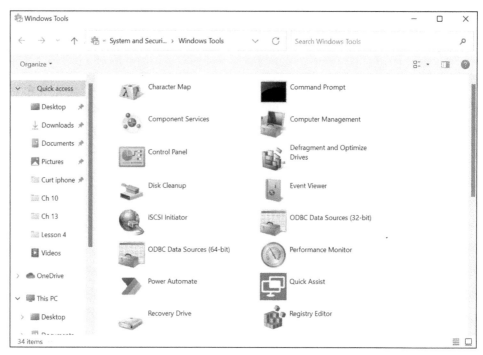

FIGURE 13-11

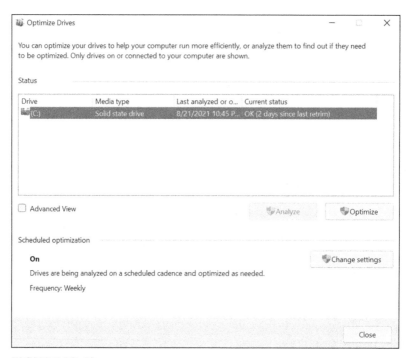

FIGURE 13-12

3. Select the C: drive and then select the Optimize button. Windows 11 optimizes your hard drive by moving the bits and pieces of files so that they are stored next to each other on the hard drive. It can take a few minutes or a few hours to optimize a hard drive, but no matter, because Windows 11 conducts this activity in the background and you can go on to do other tasks.

4. Return to the Windows Tools screen (refer to **Figure 13-11**) and select Disk Cleanup. The Disk Cleanup for Windows dialog box appears, as shown in **Figure 13-13**. It tells you how much disk space you can make available on your hard drive by removing unnecessary files.

FIGURE 13-13

5. Select the types of files you want deleted, and then select the Clean Up System Files button. A message box appears as the files are deleted from your computer.

Protect Your Computer from Viruses and Other Threats

1. Windows can identify and remove viruses, spyware, and other malefactors from your computer. To refine how Windows secures your computer against outside threats, select the Search icon on the taskbar and type **virus** in the Search box. In the Search results, choose Virus & Threat Protection. The Windows Security page opens, as shown in **Figure 13-14.** This page tells you whether the defense mechanisms on your computer are up to date. If they aren't up to date, you can take actions here to update your computer's defenses.

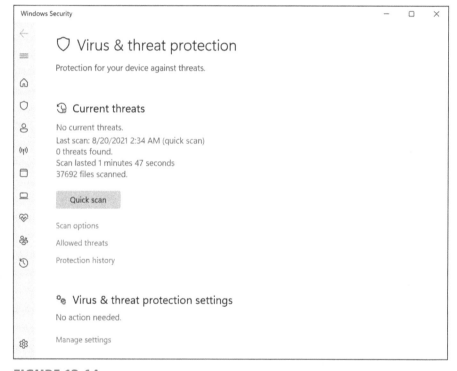

FIGURE 13-14

2. Windows performs "quick scans" in the background each time you run your computer. From time to time, to make sure your computer is thoroughly clean of viruses and spyware, run a full scan. Select the

Scan Options link to display the scan options, shown in **Figure 13-15.** Then make sure that the Full Scan option is selected and select the Scan Now button.

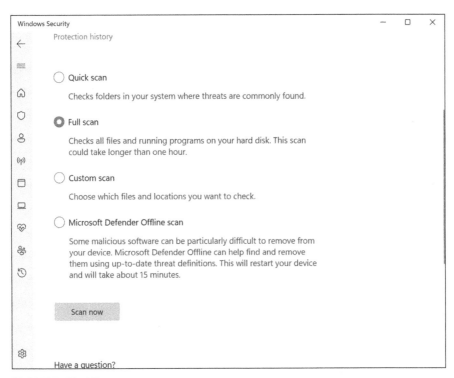

FIGURE 13-15

3. A full scan can take a long time as Windows examines all the files on your computer for viruses and spyware. You can do other tasks on your computer while the full scan is running. If Windows finds a bad actor, the file with the virus or spyware is quarantined so that it won't harm your computer.

4. Close the Windows Security window.

Chapter **14**

Connecting a Printer and Other Devices

E very computer has a screen. Most computers, other than tablets, also have a keyboard and a mouse or other pointing device. You can add a mouse to a laptop that lacks one or replace the keyboard or mouse that came with your computer. Add a printer or a scanner to extend your computer's functionality.

For any hardware add-ons — which tech folk call *peripherals* — Windows 11 has a trick up its sleeve. Thanks to *plug-and-play* technology, which automatically identifies add-on devices, connecting new devices to your computer can be quite easy.

In this chapter, you explore the devices connected to your computer, as well as options available for those devices. You find out how to access devices connected to your computer and also calibrate a touchscreen to make it work better for you.

Trust USB Plug and Play for Hardware

You may find many kinds of add-on devices useful:

» A **printer** lets you, well, print documents and photos. Your choices for printers include black and white versus color, and inkjet versus laser printer. A multifunction printer also works as a copier, scanner, and fax machine.

» A **digital camera** captures photos that you can copy to your computer to enjoy and to share with others. See Chapter 11 for information on working with photos.

» A **scanner** enables you to make digital images of old photos or documents so that you can view them onscreen. Chapter 11 explains how to use a scanner.

» An **external hard drive** stores backup copies of your files. See Chapter 16 for information on adding an external hard drive to your computer.

» An additional or replacement **pointing device** (your mouse is a pointing device), including a trackball or a pen with a tablet, may be more comfortable to use than what came with your computer. Switching between pointing devices helps you avoid repetitive stress. A wireless mouse eliminates the hassle of dealing with a cord. Some people like to add a mouse as an alternative to their laptop's built-in touchpad.

» A **microphone** is crucial for communicating by voice with your computer, through speech recognition, or with your friends over the Internet. A combination headset with microphone may produce the clearest sound.

» A **video camera** (or *webcam*) is essential for video phone calls *à la* the Jetsons. See Chapter 11 for information on using a video camera.

Some of these devices are built into computers; some are not. Most devices that aren't built in connect to a computer using *USB* (Universal Serial Bus) technology. When you connect a device to your computer using a USB cable to the USB port (see **Figure 14-1**), the device identifies itself to the computer. This identification process is called *plug-and-play.* Ideally, you connect your device, and it simply works.

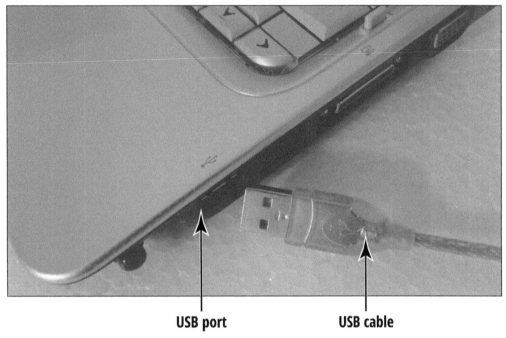

USB port   **USB cable**

FIGURE 14-1

Windows 11 uses a *device driver* to communicate with an add-on device. The driver is really a program that tells Windows 11 how to run the device. When you connect a device, such as a printer, Windows 11 looks for a driver (in this case, a *printer driver*). That driver may be built into Windows 11, come on a disc that's packaged with the device, or need to be downloaded from the Internet, either automatically by Windows 11 or manually by you.

TIP

Every computer has at least one USB port. Some are in the front; others are in the side, back, or top of the computer and are harder to reach. If your computer doesn't have enough ports, you can add more by buying a USB hub, which is a small box with several USB ports. If a port is hard to reach with a device's cable, you can buy a USB extension cable. Office supply stores may have hubs and cables, or you can easily find them online.

TIP

If you have a newer computer, it may also have one or more USB-C ports. The USB-C is a newer type of USB connection that uses a smaller connector shape than the former USB and it's easier to plug in. Also, USB-C cables are much faster than USB. Consult your computer documentation for more details.

TIP

Just because you can plug something into your computer doesn't mean you should! Flash drives and other storage media that plug into your PC can also contain viruses, so be sure to use only items from trusted sources. In other words, if you find a flash drive on the sidewalk, don't take it home and plug it into your computer!

TIP

Bluetooth is a wireless technology for adding devices to your computer. If your computer has Bluetooth, you can use Bluetooth as well as USB to add some devices, such as a microphone, headset, or wireless keyboard and mouse.

See All Devices

1. Select the Search icon on the taskbar and type **settings** in the Search box; then click the Settings app that appears in the search results. The Settings window opens. Choose Bluetooth & Devices to open the Bluetooth & Other Devices category of the Devices window, as shown in **Figure 14-2.** Devices listed may include your monitor, speakers, headphones, keyboard, mouse, and more. Devices shared through your network (if your computer is connected to a network) also appear here. For information on adjusting device settings, see the "Access Device Options" section, later in the chapter.

2. Select the Printers & Scanners category in the Devices window, to see the screen shown in **Figure 14-3.** This screen lists printers and scanners that are connected to your computer.

FIGURE 14-2

TIP

Some, but not all, devices display information below the device name. A network device may display *Offline* (not accessible) or it may display nothing if it is accessible. A printer may display *Ready* or it may display nothing if the printer isn't ready.

TIP

You are unlikely to need the Add a Bluetooth or Other Device button because most devices are added *automagically* (that's a word nerds like to use). However, if you select Add a Device, Windows 11 scans for additional hardware that you may have connected to the computer. No harm in doing so.

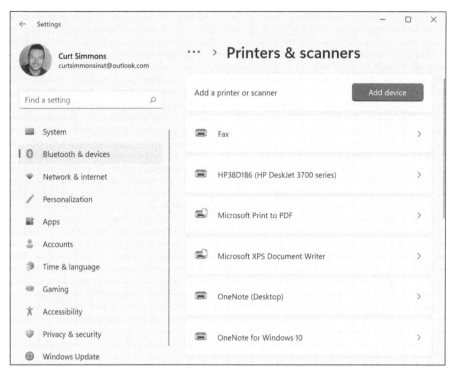

FIGURE 14-3

Connect a Printer or Other Device

1. Take your printer out of the box. Keep all the packing material together until you know you won't need to return the printer. Arrange all the components for easy access. In addition to the printer, you'll probably find ink cartridges or a toner cartridge, a power cable, and some basic installation instructions. Read the setup instructions that come with your printer before you do anything else!

TIP

Some of these steps apply to other devices, such as a mouse, a webcam, or a microphone. Printers often have more packaging and require more assembly than other devices.

TIP

Most printers connect wirelessly to your home Internet router. That way, you can wirelessly print from any computer, tablet, or even your phone. If you need to connect the printer to your computer using a USB cable, you'll need to buy one because USB cables aren't typically included with the printer. Fortunately, they are cheap and available at most big-box stores or online.

2. Remove all tape from the printer. Most printers ship with the print mechanism locked in place to prevent it from moving during shipping. Look for brightly colored tape, paper, or plastic indicating what you need to move or remove to release the print mechanism.

3. Put the printer within cable length of your computer if you're not using a wireless printer. Insert the ink or toner cartridge before you turn on the printer for the first time. Place some paper in the paper drawer or tray. Connect the printer to the power supply. Plug the printer cable into the printer and into the computer. If you're using a wireless printer, follow the setup instructions for connecting the computer to your Internet router.

TIP

Very likely, you have to download and install a software program from your printer's manufacturer in order to install your printer. Chapter 13 describes how to install programs from a company website.

4. Turn on the printer. A notification tells you that Windows is "setting up a device." If all goes well, you soon see this notification: *Device is ready,* your printer (the brand and model number of your printer) *is set up and ready to go.* You may see other informational messages as Windows 11 handles the configuration, and you may see messages from the printer software as well. Just follow the printer manufacturer's instructions as needed.

5. To confirm that your printer is installed properly, see the preceding section, "See All Devices."

Access Device Options

1. For more control over device setup, open the Control Panel to the Devices and Printers window. To do so, select the Search icon on the taskbar and type **control panel** in the Search box. Select Control Panel in the search results to open the Control Panel, and select View Devices and Printers (under Hardware and Sound) to display the window shown in **Figure 14-4.**

2. The Device and Printers window shows the devices attached to your computer, including the computer itself, the display (or monitor), external add-on devices (such as a hard drive, flash drive, or memory card), and the mouse.

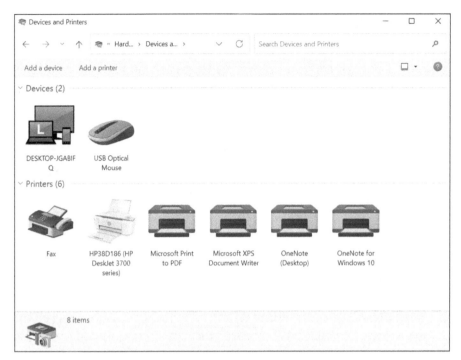

FIGURE 14-4

TIP

Most of these devices also appear in the Devices category of the Settings app (refer to **Figure 14-3**). However, you'll find options under Devices and Printers that aren't available in the Settings app.

TIP

Windows 11 automatically installs the Microsoft Print to PDF software. This software doesn't print but does create files that you can print later using a real printer. For example, if you're in a coffee shop and want to print a web page or an email message, you can use this device and then open the file it creates when your computer is connected to a printer.

3. Double-click or double-tap the device you want to examine. This action opens the device's properties in a window with options or in a smaller box with limited information and options. (Older devices have more limited information.) **Figure 14-5** shows information about a printer. When you're finished reviewing the information or selecting available options, return to the previous screen by clicking the back arrow on the toolbar.

FIGURE 14-5

4. Display the context menu of options for your printer (or any device) using one of these methods: Click the right mouse button, or tap and hold until you see a small box appear under your finger, and then release. Select Printer Properties from the context menu. (Oddly, the menu also has a separate Properties option — be sure to select Printer Properties instead.)

5. In the Properties window, select the Print Test Page button. Another window opens indicating *A test page has been sent to your printer.* Select Close.

TIP

If a test page doesn't print, check that both ends of the cable are plugged in properly and make sure the printer is turned on. If you're using a wireless printer, make sure the printer is connected to your Internet router. Try to print a test page again. For more help, contact the printer manufacturer or the store where you bought the printer, or search the web.

TIP

If you're having problems with any device, select the Trouble-shoot option on the context menu to open a guided troubleshooting program that will walk you through options for resolving problems with the device.

TIP

The top of the Devices and Printers window has the Add a Device and Add a Printer buttons, but you need to use them only if Windows 11 doesn't automatically detect and install your device. With USB and plug-and-play technology, most devices install automatically.

Calibrate Your Touchscreen

1. If you have a problem accurately selecting objects on your screen using touch, you can calibrate your screen alignment. In this case, calibrating means to help Windows 11 understand what constitutes a tap on the screen. Select the Search icon on the taskbar and type **control panel** in the Search box. Select Control Panel in the search results and then select Hardware and Sound in the Control Panel window. You see the Hardware and Sound window, shown in **Figure 14-6.**

2. In the Hardware and Sound window, select Calibrate the Screen for Pen or Touch Input (it's under Tablet PC Settings); see **Figure 14-6.** The Tablet PC dialog box appears (see **Figure 14-7**).

3. Select the Calibrate button. User Account Control may ask you to confirm that you want to run the Digitizer Calibration Tool. If so, select Yes.

TIP

If your touchscreen is badly calibrated, you may not be able to tap the Calibrate button. In that case, plug in a mouse to make the selection, and then continue using touch.

4. The screen displays lines around its perimeter, forming a box near the edge of the screen and a second box inside the first about half an inch from the edge. Lines connect these boxes near each corner. The result is 16 intersections. Starting at the upper-left corner, use your finger or a stylus to tap each intersection, which displays two short black lines forming crosshairs. As you touch each intersection,

Windows 11 measures your touch and adjusts touch settings accordingly. After each touch is recorded, the crosshairs move to the next intersection to the right. (If the crosshairs don't move, tap the previous intersection again.) As the crosshairs move, tap the highlighted intersection, left to right, down, and then left to right again. The process takes much longer to read about than to do.

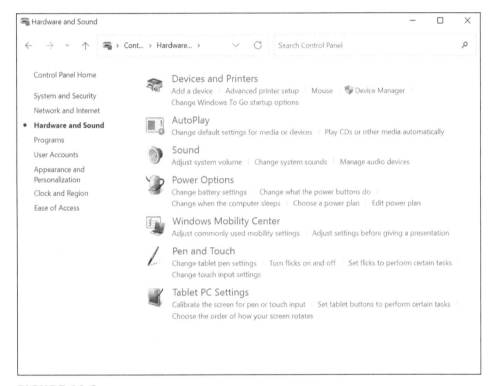

FIGURE 14-6

TIP

If Windows 11 doesn't recognize your touch, it won't continue with the process. The screen says *Right-click anywhere to return to the last calibration point*. What if you don't have a mouse? Tap and hold until you see the little box under your fingertip, and then release — that's the touch equivalent of a click of the right mouse button.

5. After you have selected each of the 16 calibration points in turn, a box pops up asking whether you want to save the calibration data. Select OK unless you think something went wrong. In that case, select Cancel.

FIGURE 14-7

Chapter **15**

Working with Files

ll the data inside your computer is stored on a disk. Your computer has a primary disk, formally called the internal *hard drive.* You may see this disk referred to as the C: drive. (The terms *drive* and *disk* are interchangeable.)

The contents of a disk are organized into individual files. When you create a file and save it for the first time, you create a file on a disk. Many other files on the disk belong to the programs you use, including the thousands of files that make up Windows 11.

Disks also are organized into *folders,* which are containers for files. For its own files, Windows 11 has a main folder that contains dozens of other folders (called *subfolders*). Inside or below that user account folder, Windows 11 creates more folders to help you organize your files by type. For example, by default, all your photos go into the Pictures folder, and all your documents go into the Documents folder.

In this chapter, you discover how to save and open a file, search for files, and explore your disk, folders, and files. You work with File Explorer as you create new folders to organize documents and move files from one folder to another. You also copy files from your hard drive to other disks to take with you or give to other people. This chapter also explains how to rename and delete files and folders. What if you delete a file or folder and you regret doing so? Don't worry, because this chapter shows how to get back a file or folder that you mistakenly deleted.

Save and Open a File

1. In computer land, nothing is more important than files. Computer programs are stored in files. Data is stored in files. To start working, you open a file. When you're finished working, you save and close your file. To experiment with files, open WordPad, the word processing app that comes with Windows 11.

TIP

Chapter 2 explains how to open apps. Chapter 9 looks into WordPad and how it can serve your word processing needs.

2. As shown in **Figure 15-1,** type a few words into WordPad and then save your file. Most apps (WordPad included) offer three ways to save a file:

- Select the Save button.
- Press Ctrl+S.
- Open the File menu and choose Save.

3. When you save a file for the first time, the Save As dialog box appears (refer to **Figure 15-1**). It invites you to give the file a name and to tell Windows 11 which folder to keep it in. Enter the name **My File** in the File Name text box. The Save As dialog box offers the means of choosing a folder for your file. For now, you can store your file in the Documents folder, the folder that Windows 11 selected for you (refer to **Figure 15-1**). Notice that the Documents folder is already selected in the navigation pane on the left side of the Save As dialog box.

4. Select the Save button in the Save As dialog box. Congratulations! You just saved your file. The title bar at the top of the WordPad window shows the name of your file, "My File." When you're not sure what a file is called, look at the title bar along the top of the screen.

5. Close WordPad. The fastest way to close an app is to select the Close button (the *X*) in the upper-right corner of the screen.

6. Open WordPad as you did in Step 1. Where is My File, the file you saved and named in Steps 4 and 5?

7. Open the File menu (select the word *File* in the upper-left corner of the screen) and choose Open. The Open dialog box appears, as shown in **Figure 15-2**. This dialog box is the mirror image of the Save As dialog box that you saw earlier. Use it to locate the file you want to open. In this case, Windows 11 has opened the dialog box to its favorite folder, Documents, which also happens to be the folder where My File is stored.

8. Choose My File and then select the Open button (refer to **Figure 15-2**). My File opens onscreen.

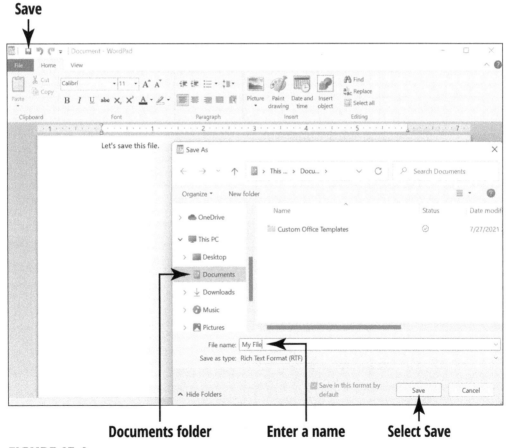

FIGURE 15-1

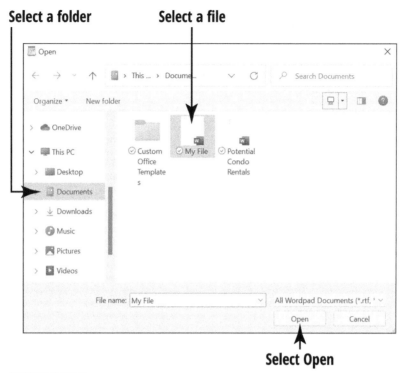

Select a folder   **Select a file**

Select Open

FIGURE 15-2

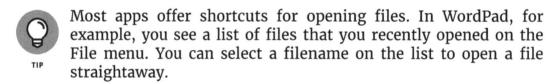

Most apps offer shortcuts for opening files. In WordPad, for example, you see a list of files that you recently opened on the File menu. You can select a filename on the list to open a file straightaway.

TIP

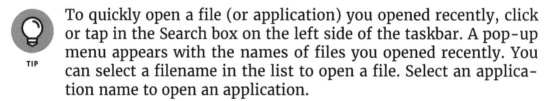

To quickly open a file (or application) you opened recently, click or tap in the Search box on the left side of the taskbar. A pop-up menu appears with the names of files you opened recently. You can select a filename in the list to open a file. Select an application name to open an application.

TIP

Find a Misplaced File

1. To search for a misplaced file, begin with the Search icon, located on the taskbar. Type the name of a document or photo you have on your computer. The search results appear, as shown in **Figure 15-3.**

Narrow the search   **Enter a search term**

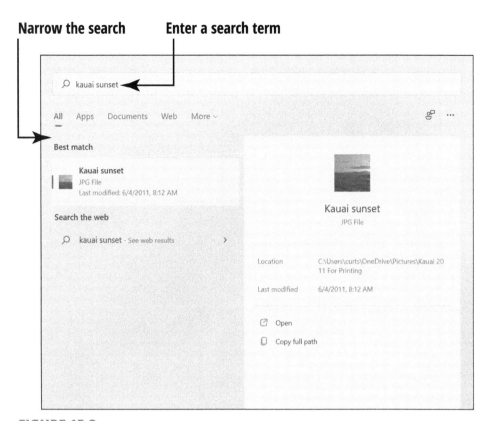

FIGURE 15-3

TIP

Initially, the search results shows all and sundry — settings, apps, files, websites, and other search opportunities — in the search results. If you don't see the item you want in a search, narrow the list of items by selecting one of the tabs at the top of the search window — Apps, Documents, Web, or More. Documents, for example, narrows the search to items on your computer.

2. To open a file, select it in the search results. If more than one application on your computer is capable of opening the file, you may see the How Do You Want to Open This File? dialog box, as shown in **Figure 15-4.** Select an application to open the file.

TIP

The How Do You Want to Open This File? dialog box offers a means of telling Windows 11 to always open files of a certain type with a particular application (refer to **Figure 15-4**). Selecting the Always Use This App to Open check box tells Windows 11 to open files of a certain type automatically with the app you chose.

Choose an application

How do you want to open this file?

Keep using this app

Photos
Photos are easily enhanced, organized for you, and
ready to edit and share.

Other options

Adobe Photoshop Express
New

Paint

Paint 3D

Snip & Sketch

☐ Always use this app to open .JPG files

OK

**Select to always open files of this type
with the application you chose**

FIGURE 15-4

3. Close the app that opens the file by using one of these methods:

- Click or tap the Close button (the X located in the upper-right corner of the screen).

- Press Alt+F4.

TIP

Strictly speaking, if you find what you're looking for, stop looking. (No extra charge for pearls of wisdom.) In this case, however, keep looking, because doing so reveals important information about how files are organized and how you can take control of that organization.

4. On the taskbar, select the File Explorer icon (the yellow folder). File Explorer opens. Notice that File Explorer contains a toolbar, a left navigation pane, and a right content pane that displays the contents of the selected folder, as shown in **Figure 15-5.**

Toolbar

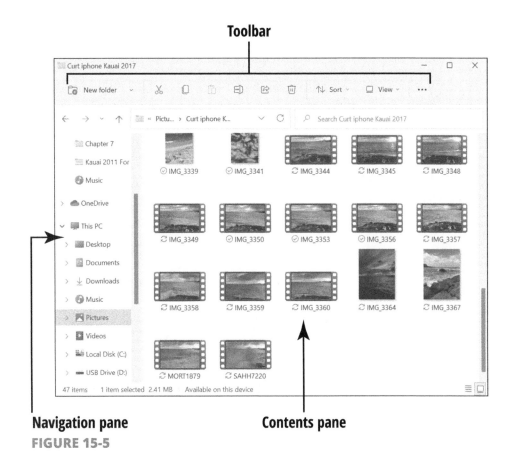

Navigation pane   **Contents pane**

FIGURE 15-5

5. Select This PC on the left side of the File Explorer window, and then select the box labeled *Search* (probably followed by *This PC*), below the toolbar and to the right. Type the same search term you used in Step 1. As you type, File Explorer displays any matching files, highlighting the text that matches. **Figure 15-6** shows the results of a search on a computer using the search term *Kauai sunset*.

TIP

On a touchscreen, the virtual keyboard doesn't appear on the desktop until you select the keyboard icon on the right side of the taskbar.

TIP

If the search results include too many files, thereby making it hard to see the one you want, type more of the filename in the Search box. The number of matching files should decrease as you type more text in the box.

6. The *focus* of a search — where File Explorer searches — determines what files and folders are found. In Step 5, you selected This PC, making all the files on your computer the focus of the search. Accordingly, Windows 11 searched for files and folders throughout your computer. To change the focus of your search to the Pictures folder, select the Pictures folder in the navigation pane. Then select the box labeled Search and enter your search term again. The search results focus on files in the Pictures folder.

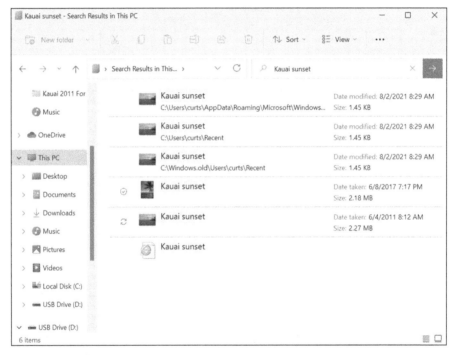

FIGURE 15-6

TIP

You can use the Sort feature on the toolbar to refine a search, as needed. Start a search and then click Sort⇨ More. You can then choose to sort by date created, authors, tags, and so forth.

Add a Location to Quick Access

1. If you frequently access a specific location using File Explorer, you might want to add that location to the Quick Access section of the navigation pane for easy access. The Quick Access section is located at the top of the navigation pane. Windows 11 places some folders in the Quick Access folder list automatically, including the Documents folder and the Pictures folder. As well, it lists in alphabetical order folders that you visit frequently.

2. Display the Quick Access folders. To do so, click or tap the arrow next to Quick Access in the navigation pane to expand it, as shown in **Figure 15-7.**

Click the arrow to expand

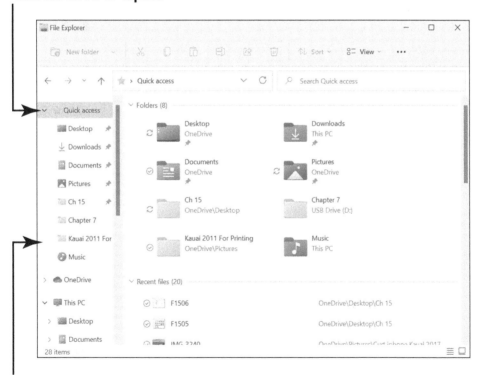

Quick Access folders and files

FIGURE 15-7

3. Select a folder that you access frequently. For example, if you're currently working on files you keep in a folder called My Work, select that folder in the navigation pane.

4. On the See More button on the toolbar (it looks like three dots), select the Pin to Quick Access option on the menu (refer to **Figure 15-8**). The name of the folder you selected in Step 3 now appears in the Quick Access folder list. Notice the pin beside its name.

5. Select your folder in the Quick Access folder list. Next time you want to get to this folder, you can get there quickly by selecting Quick Access in the navigation pane and then selecting your folder.

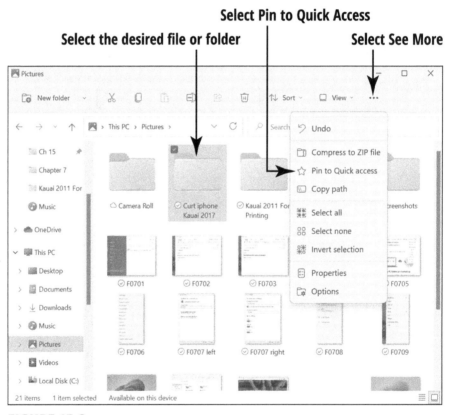

FIGURE 15-8

TIP

To remove a folder from the Quick Access list, select the folder you want to remove. Then right-click the folder and choose Unpin from Quick Access on the shortcut menu.

Use File Explorer for Easy Access to Files

1. Select the File Explorer icon on the taskbar if File Explorer is not already open. The right side of Explorer is called the *content area.* The left side of Explorer is called the *navigation pane* and contains folders. Explorer starts with its focus on your Quick Access and This PC folders, and on your C: drive and Network drives (if your computer is connected to a network).

TIP

The keyboard shortcut to open File Explorer is ⊞+E.

2. On the left, select Music, and then Pictures, and then Videos, while noting the files in the content area on the right. Many Windows 11 apps store files by default in the Documents, Music, Pictures, and Videos folders. Select the Documents folder again.

3. To create a practice document that you can use in later sections, select the New Folder menu on the toolbar. In the New section, select Text Document, as shown in **Figure 15-9.** (If you don't see this option, make sure that the Documents folder is selected.) An empty text document is created and the words *New Text Document* are highlighted so that you can type a new name. Type **practice file**. (You'll rename this file in a later section.) Feel free to repeat this step to create additional items for practice, such as Microsoft Word Documents or Bitmap Images.

TIP

On a touchscreen, the virtual keyboard doesn't appear on the desktop until you select the keyboard icon on the right side of the taskbar.

4. Select the View button on the taskbar. In the drop-down menu that appears, select each option, such as Extra Large Icons and Large, Medium, and Small Icons. Try them all, if you like.

Point to New

Select New Folder

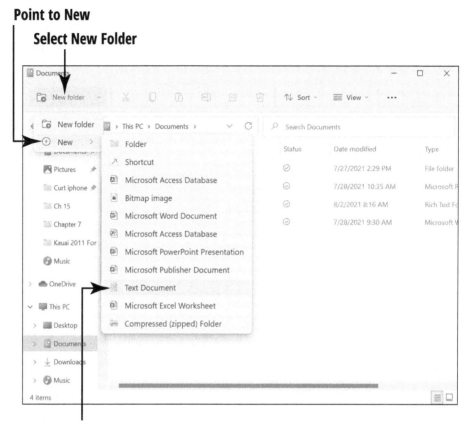

Select Text Document

FIGURE 15-9

TIP

Certain layouts are better for certain purposes. For example, photos are easier to recognize as Extra Large Icons than as a List. **Figure 15-10** shows documents using the Details view, which includes the date the file was modified.

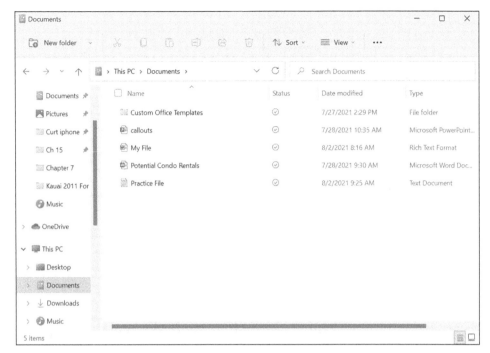

FIGURE 15-10

Create a Folder to Organize Your Files

1. In the File Explorer navigation pane, select the Documents folder.

2. Select the New Folder button on the taskbar. An icon for the new folder appears in the content area on the right, with the name *New folder* next to it and already selected (see **Figure 15-11**).

TIP

On a touchscreen, the virtual keyboard doesn't appear on the desktop until you select the keyboard icon on the right side of the taskbar.

3. Type **practice folder** as the new name. Don't move the cursor or mouse before you start typing. Your new text will replace the highlighted text automatically. Press the Enter key to make the new name stick. (If you have a problem naming the folder, see the "Rename a File or a Folder" section, later in this chapter.)

4. Open your new folder by double-clicking or double-tapping its icon. Notice that the content area is empty.

5. To return to the Documents folder, select Documents in the navigation pane.

Select New Folder   **The New Folder appears**

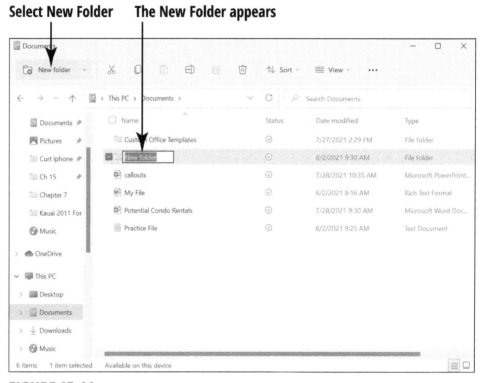

FIGURE 15-11

TIP

Don't worry too much about creating folders, because the folders that Windows 11 provides may be all you ever need. As you accumulate more files, however, placing them into other folders can help you stay organized. In the Documents folder, for example, you might create a subfolder called Finances for files related to income, expenses, and investments, and another subfolder called Family for family-related documents. Which folders and subfolders to create and how to name them depends entirely on your own sense of order.

Use Check Boxes to Select Files

1. In File Explorer, you select files to move, copy, rename, or delete. You can add a check box to make selecting multiple files easier. This option may be turned on by default. Select a file or folder in File Explorer, and if a check box appears next to it, you don't need to do anything else. If not, however, you can turn on this feature. In File Explorer, select the View button on the toolbar and then select Show. Click the Item Check Boxes option on the menu that appears, as shown in **Figure 15-12.**

2. Select the Documents folder (or any folder that contains more than one file). To select a file, click or tap its filename. (You won't see the check box until you select it with a tap or a click, or hover over it with the mouse pointer.) Repeat to select additional files. If you want to deselect a file, select the check box again to remove the check mark. Close the window after you've seen how these check boxes work. **Figure 15-13** shows four selected files.

TIP

If you want to select only a single file, you can select anywhere on the filename. You use the check box when you want to select more than one file at a time.

The See More button on the toolbar offers other methods for selecting. Select All does just what it says — selects all objects in a folder or library. Select None works similarly. Invert Selection switches the selection. For example, if we chose Invert Selection in the context of **Figure 15-13,** the four selected files would be deselected and all the other files would be selected. Sometimes it's easier to select the files you don't want and then invert the selection.

Select View

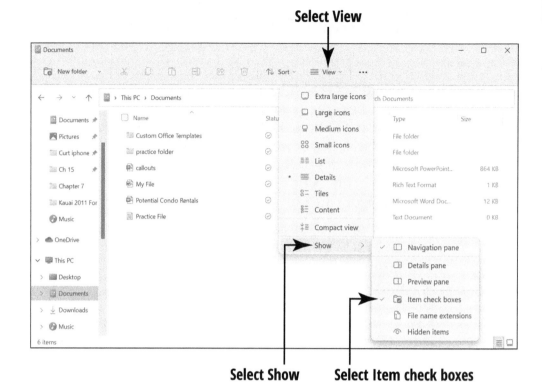

Select Show   Select Item check boxes

FIGURE 15-12

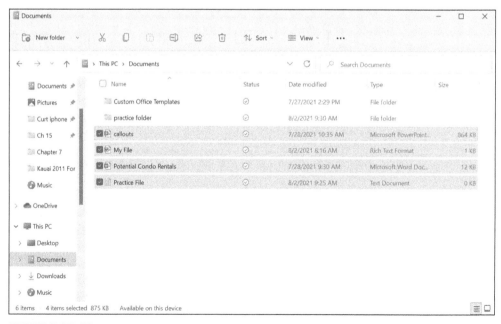

FIGURE 15-13

Undo or Redo an Action

1. Sometimes you make mistakes when you're working with Windows 11. For example, you may give a file the wrong name or move a file to the wrong folder. The good news is that you can easily undo your last action. Just click the See More button on the toolbar and you see Undo in the menu list, as shown in **Figure 15-14.**

2. If you make a mistake and want to undo your last action, just click the Undo button. Note that there is no time limit for Undo, so no rush. However, also note that Undo undoes only your last action. You can't use this feature to undo something you did several steps back as you're working in File Explorer.

The keyboard shortcut for undo is Ctrl+Z.

3. You can also redo an action that you've chosen to undo. Say you undo a filename for a document, but then you realize that the original file name was correct after all. In that case, just select See More and then select Redo, as shown in **Figure 15-14.**

The keyboard shortcut for redo is Ctrl+Y.

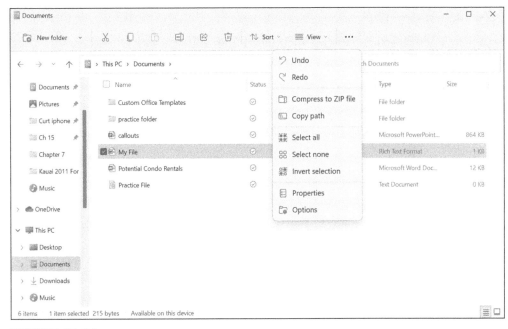

FIGURE 15-14

Move a File from One Folder to Another

1. You can move files to organize them. For this exercise, select the Documents folder in File Explorer. Select one of your documents.

To move more than one file at a time, see the section "Use Check Boxes to Select Files," earlier in this chapter.

TIP

2. In the navigation pane, scroll to find the folder you want to move the file to. When you see it, just drag the file to the new folder in the navigation pane and release it. This action moves the file out of the original folder and to the new folder you've chosen.

Use these same steps to move a subfolder from one folder to another. However, don't move folders that Windows 11 creates.

TIP

Rename a File or a Folder

1. You can change the name of any file or folder you create. (Don't rename files in the Windows or Program Files folders.) For this exercise, select the Documents folder in File Explorer. Then select one of your files.

To rename more than one file at a time, see the section "Use Check Boxes to Select Files," earlier in this chapter. On completion of the rename operation, the files you selected will share the name you provide; each file will have a unique number added to the name, starting with *(1)*.

TIP

2. On the toolbar, select the Rename button. In the content area, the current name of the file or folder is selected, as shown in **Figure 15-15.** If you type anything while the text is selected, you erase the current name, which is convenient if the new name is completely different from the old name. If you want to keep most of the current name and edit it, select inside the name or press the left- or right-arrow key to move to the place in the name where you want to type new text.

3. Type the new name, which can be more than 200 characters long (although a dozen characters may be more than enough). You can capitalize letters and use spaces and dashes, but you can't use slashes or asterisks, which Windows 11 reserves for other purposes.

TIP

On a touchscreen, the virtual keyboard doesn't appear on the desktop until you select the keyboard icon on the right side of the taskbar.

4. When you've typed the new name, press the Enter key to finish the process.

Type a new name for the file

Select Rename

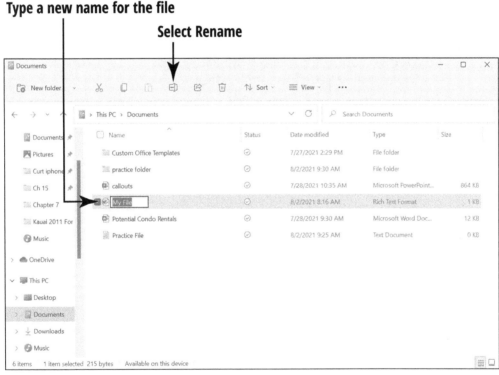

FIGURE 15-15

TIP

You can right-click a file's icon to see the Rename option on the context menu that appears. You may find this method faster.

Delete a File or Folder

1. You can delete any of your files that you no longer need. (Don't delete files in the Windows or Program Files folders.) For this exercise, select the Documents folder in File Explorer. Then select one of your files.

   TIP

   To delete more than one file at a time, see the section "Use Check Boxes to Select Files," earlier in this chapter.

2. On the toolbar, select the Delete button.

   TIP

   The keyboard shortcut to delete the selected file is the Delete key (surprise!).

3. A confirmation dialog box appears and asks whether you really want to delete the file, as shown in **Figure 15-16.** Here's your chance to change your mind if you want. However, for this exercise, select Yes in the dialog box to delete the file.

   TIP

   If you don't see a confirmation dialog box like the one in **Figure 15-16,** tell Windows 11 that you want to see this dialog box when you delete files and folders. Just right-click the Recycle Bin and click Properties. On the Properties dialog box, select Display Delete Confirmation Dialog and click OK, as shown in **Figure 15-17.**

4. Select another file, and then select the Delete button. The Delete File confirmation window appears. This time, select No to cancel the operation.

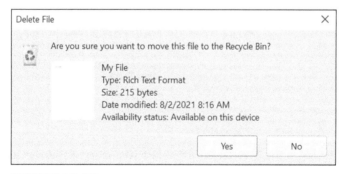

FIGURE 15-16

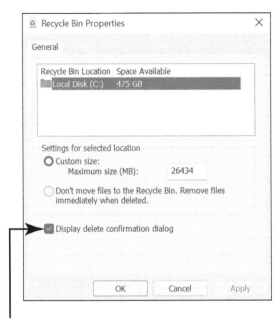

Select this check box and click OK

FIGURE 15-17

Get Back a File or Folder You Deleted

1. Normally, when you delete a file or folder, Windows 11 moves the object to the Recycle Bin. Objects remain in the Recycle Bin indefinitely, allowing you to restore something you deleted long after you did so. To open the Recycle Bin, go to the desktop and double-click or double-tap the Recycle Bin icon. The Recycle Bin opens, as shown in **Figure 15-18.**

2. If many files or folders are listed in the Recycle Bin window, type the name of the item you want in the Search box in the top-right corner of the window. If any files match what you type, they appear in the content area.

Note that on a touchscreen, the virtual keyboard doesn't appear on the desktop until you select the keyboard icon on the right side of the taskbar.

3. To restore a file or folder to its original location, select the file or folder in the Recycle Bin window. Select See More on the toolbar and select Restore the Selected Items. The selected file or folder returns to the folder it was in before it was deleted.

TIP

If Windows 11 needs disk space, it will automatically clear out the oldest files in the Recycle Bin first. If you want to get rid of everything in the Recycle Bin, select Empty Recycle Bin on the toolbar. After you empty the Recycle Bin, you can't undo your action.

TIP

Don't select the Restore All Items button, because doing so puts every single item in the Recycle Bin back in its original location. Most of the files in the Recycle Bin are probably files that you really meant to delete. Choosing this command would be like dumping the trash can on your living-room floor to find a penny you threw away.

See Chapter 16 for information on backing up and restoring files.

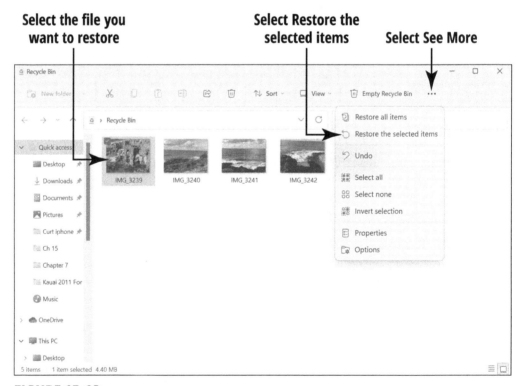

FIGURE 15-18

IN THIS CHAPTER

» **Attaching files and copying them to a storage drive**

» **Managing files on OneDrive**

» **Creating and storing backup files**

» **Restoring a failed computer**

Chapter **16**

Backing Up and Restoring Files

Some of your files — photos and documents — are priceless. If you accidentally delete a treasured file, what can you do but cry? You can insure your well-being by creating copies of your documents and photos.

The best insurance involves storing copies of files on devices separate from your computer. Such devices include the following:

» **Flash drive and memory card:** Carry your files when you're away from your computer by storing them on a portable storage device. For example, you can store files on a USB *flash drive* (also called a *thumb drive*), or a *memory card,* which is the size of a postage stamp and is most often used in laptop computers and digital cameras. Common capacities for flash drives and memory cards range from 8 to 256GB.

TIP

A *gigabyte* (GB) of storage can hold thousands of files, but you'll be amazed by how quickly you can fill that space.

» **External hard drive:** This type of drive has a much higher capacity than a flash drive, making it ideal for backing up all your files — the best insurance. Affordable external hard drives range from 500GB to 3TB.

TIP

A *terabyte* (TB) of storage is equal to a thousand gigabytes, which should be enough room to back up everything on your computer.

» **Network drive:** If your computer connects to a home network, you may be able to copy files to other devices on the network. For example, you can use a large-capacity network drive to back up files from more than one computer.

» **OneDrive:** Your Microsoft Account comes with 5 gigabytes of free storage in the cloud (which means on the Internet). Anything stored in OneDrive is duplicated on additional computers you log in to with the same Microsoft Account.

TIP

OneDrive provides storage but not a backup (duplicate). If you delete a file from OneDrive, any copies stored on linked computers are also deleted.

In this chapter, you copy files to a flash drive and hard drive so that you can transport files between machines and back up your files. You also find out how to store files in the Microsoft cloud with OneDrive. As well, you use the Backup function, which automatically copies files as a backup. Consider this scenario: You write a letter to a friend and save it to your Documents folder. Later that day, you delete part of the letter and save it again, replacing the original document. The next day, you wish you still had the deleted text. The Backup function comes to your rescue because it saves versions of files; you can recover the latest version or an earlier version of a file. As we wrote this book, we saved it hundreds of times — the Backup function could save every version, allowing us to roll back to an earlier copy, to before we had made some big goof. (That's purely hypothetical, of course.)

Finally, in this chapter, you explore the Reset function, a tool you may need if you have problems with your computer. The Reset function reinstalls Windows 11 but preserves your personal data.

Add an External Hard Drive or Flash Drive

1. Before you attach a flash drive or hard drive to your computer, consider the following options that Windows 11 automatically offers for using the newly attached drive:

   - **Configure Storage Settings** opens the Settings app so that you can configure the storage settings on your computer.

   - **Open Folder to View Files** displays the contents of the disk in File Explorer on the desktop. You select this option to copy files to or from the drive you're attaching.

   - **Take No Action** dismisses the notification.

TIP Notifications appear for the amount of time specified in PC Settings. If the notification disappears before you can select it, you can redisplay it by removing and then reinserting the drive or USB cable.

2. Locate an unused USB port on your computer. A *USB port* is a small rectangular slot on the front or back of a desktop computer or along any edge of a laptop or tablet computer. USB ports are often marked with a symbol that looks like a trident, as shown in **Figure 16-1.**

TIP If a USB port is hard to reach, you can buy an extension cable from any office supply store. You can also buy a *hub,* which adds ports to your computer.

3. If you're using a flash drive, insert it into the USB slot. USB fits one way only. If you're using an external hard drive, plug it into a power source, if one is required, and then connect a cable to the USB port. Turn on the external drive, if it has a separate power switch. (Flash drives and some external hard drives don't have separate power supplies or switches.)

4. Windows 11 displays a notification to *Choose what to do with removable drives*. (You can click the mouse instead of tapping.) If you select the first notification, a list of choices appears, such as *Open folder to view files*. If you know which action you want to take, you can select that action. Otherwise, select Take No Action or wait until the notification disappears on its own.

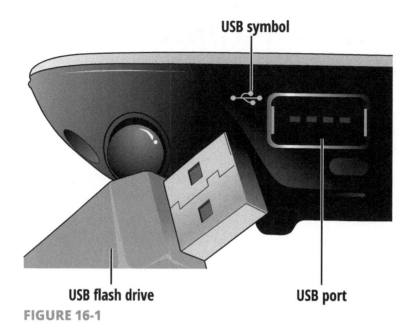

USB symbol

USB flash drive   USB port

FIGURE 16-1

Copy Files to or from a Flash Drive

1. To copy files or folders to a flash drive, insert the flash drive into one of your computer's USB ports.

TIP

Laptops, like cameras, often have a slot for a memory card. Want to turn your memory card into a flash drive? Simply buy a device called a dedicated or single-purpose memory card reader. Strictly speaking, a multipurpose card reader also works, but multicard readers cost more and are often larger than single-card readers. In addition, a dedicated memory card reader doesn't need a cable to connect a camera to a computer. You can just download your pictures to your hard drive from the card.

2. If Windows 11 displays a notification, when you insert the flash drive or memory card, select Open Folder to View Files, which will open File Explorer on the desktop. If File Explorer doesn't open automatically, go to the desktop and then select the yellow folder icon in the taskbar to open File Explorer.

TIP

The keyboard shortcut to open File Explorer is ⊞ + E.

3. In File Explorer, navigate on the left to the folder that contains the files you want to copy. See Chapter 15 for information on navigating in File Explorer. Select the folder.

4. On the right side of File Explorer, select the folder or file you want to copy. If you see a check box to the left of each object you want to copy, you can select each check box to copy multiple files simultaneously. (If you don't see check boxes next to files, see Chapter 15 for information on enabling this function for file selection.)

To select every object on the right simultaneously, use the Select All button available on the See More menu.

The keyboard shortcut to select all files in File Explorer is Ctrl + A.

5. Select the Copy button, as shown in **Figure 16-2.**

6. Click the little arrow next to This PC in the File Explorer navigation pane and locate the flash drive or memory card. The drive will not be Local Disk (C:), where Windows 11 resides. Select the removable flash drive or memory card to which you want to copy the files, as shown in **Figure 16-3,** and then select the Paste button on the toolbar. If the files copy quickly, you may not see any indication of progress; otherwise, a progress bar is displayed until copying is complete.

OneDrive may be listed in the Copy Items window. Files you copy to OneDrive are automatically copied to the cloud and to linked computers.

7. If you copy a file that is already on the destination disk, the Replace or Skip Files window appears, as shown in **Figure 16-4.** (Perhaps you're copying a newer version of a file you copied before.) Note the available options:

- **Replace the File in the Destination:** Selecting this option replaces one file with another. Be certain that you don't need the replaced file (as you might if you want to keep different versions of files).

- **Skip This File:** Selecting this option does nothing with this file.

- **Compare Info for Both Files:** Selecting this option opens another window in which you can select files on the left to replace those on the right, and select files on the right to keep. Selecting the same file on the left and right creates a second file with a number added to the name, such as *my file (2)*. This option enables you to have the original and the new file.

  Select one of the previous options. If you selected Compare Info for Both Files, select the files to replace or skip, and then select the Continue button. You may or may not see a progress indicator, depending on how quickly the files are copied.

8. Confirm that the copy worked by navigating on the left to the location you selected as the destination in Step 6. If the files are there, congratulations; you're done. If not, try Steps 4 through 6 again.

9. Remove the flash drive or memory card you inserted in Step 1. You're good to go.

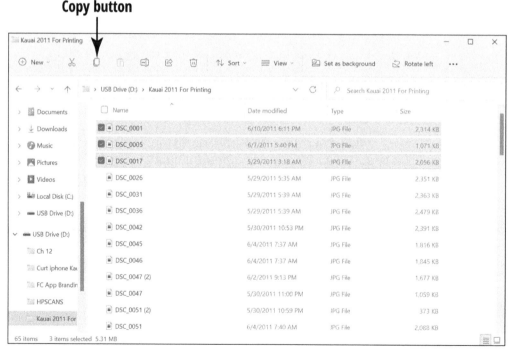

Copy button

FIGURE 16-2

Paste button

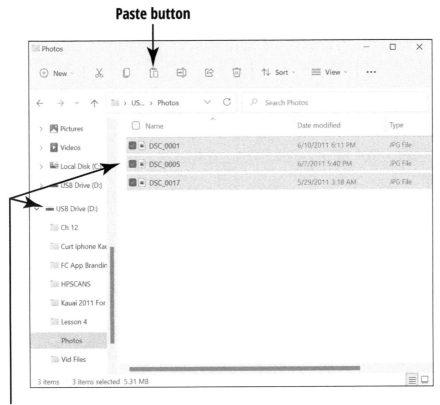

Pasted files on the USB Drive

FIGURE 16-3

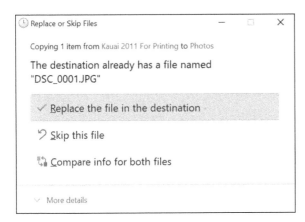

FIGURE 16-4

TIP

If you have files or folders that you'd be devastated to lose, follow the steps in this task to create backup copies of those items on a portable storage device. Then keep that device in a safe place.

TIP

To copy files from a flash drive or memory card, follow these same steps but select the flash drive in Step 3 and the folder or other destination to which you want to copy or move files in Step 6.

Use OneDrive to Store File Copies

1. Use OneDrive to back up and store files on the Microsoft cloud — in other words, to back up and store files on servers that Microsoft maintains on the Internet. If you have a Microsoft Account, you are entitled to 5 gigabytes of free storage space on the Microsoft cloud. If you have an Office 365 subscription, you are entitled to 1,054 gigabytes (1 terabyte). Here's the big advantage of OneDrive: You can access files that you keep on the cloud from any computer connected to the Internet, whether you're in Tallahassee or Timbuktu. Use one of these techniques to open the Set Up OneDrive dialog box and tell Windows 11 that you want to use OneDrive:

- Select the OneDrive icon in the notification area located in the lower-right corner of the screen.

- Select the Search icon on the taskbar and enter **onedrive** in the Search box; then select the OneDrive app in the search results.

TIP

If you're signed in with a Microsoft Account, Windows 11 already knows your password. Skip to Step 3.

2. In the Set Up OneDrive dialog box, enter your email address and select the Sign In button. In the dialog box that appears, enter the password to your email account and select the Sign In button.

3. The next dialog box welcomes you to OneDrive or tells you where your OneDrive folder is located (which dialog box you see depends on whether you are signed in with a Microsoft or a local account). Click the Next button or Close button as necessary. You're ready to copy the folders you want to back up or store into your OneDrive folder in the Microsoft cloud.

4. Open File Explorer and copy the folders you want to back up or store into the OneDrive folder. As shown in **Figure 16-5,** this folder appears in the navigation pane on the left side of the File Explorer window. Don't let appearances fool you. Although the OneDrive folder appears alongside the other folders in the navigation pane, that folder lives in the Microsoft cloud, not on your computer. Folders and files on OneDrive show a green check mark to let you know they're on OneDrive (refer to **Figure 16-5**).

TIP

Chapter 15 explains how to copy and move files with File Explorer. As well, it explains how to create and save files.

5. In File Explorer, open one of the folders on your computer that you copied to OneDrive, create a new file, and save the new file under the name **Sync** in the folder you opened.

6. In File Explorer, select the OneDrive folder and then, in OneDrive, open the folder with the same name as the folder you opened in Step 5. You see the Sync file on OneDrive. When you save a file in a folder that is also kept on OneDrive, a copy of the file is made to OneDrive automatically. Microsoft calls this *syncing* (for synchronizing). Thanks to syncing, you don't have to copy files to OneDrive on your own because the copies are made automatically.

TIP

To see which folders from your computer are synced to folders in OneDrive, select the OneDrive icon in the notification area, choose Help & Settings on the dialog box, and then choose Settings in the pop-out menu. The Microsoft OneDrive dialog box appears. On the Account tab, select the Choose Folders button. The Choose Folders dialog box appears, as shown in **Figure 16-6.** You can deselect a folder in this dialog box to prevent OneDrive from syncing a folder on your computer to its counterpart folder on OneDrive. Deselect a folder if you don't care to back up its contents on OneDrive.

7. Start the Edge browser and go to the OneDrive online website at this address: onedrive.live.com. If necessary, enter your password. You see the folders from your computer that you store on OneDrive, as shown in **Figure 16-7.** You can manage your files and folders starting here. For example, you can delete, move, and download files and folders.

OneDrive Folder

Check marks indicate files and folders are stored on OneDrive

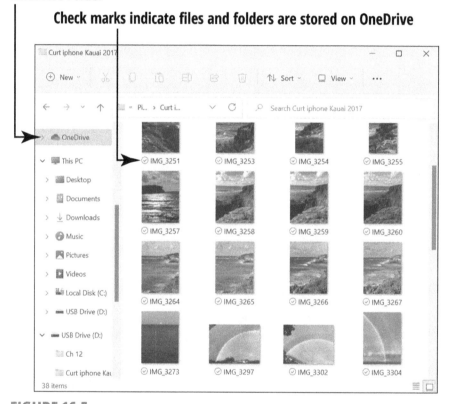

FIGURE 16-5

Chapter 6 shows how to use Edge, the official web browser of Windows 11.

TIP

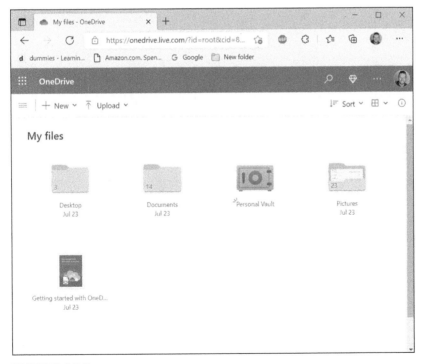

FIGURE 16-6

FIGURE 16-7

Turn On File History

1. To enable Windows 11 to create backup copies of your files, first make sure that an external drive such as a hard drive or flash drive is connected to your computer. Then search for File History and select the File History option from the search results. Ideally, you have an external drive that has as much or more storage space as you computer's internal disk drive.

2. The File History window opens, as shown in **Figure 16-8.** File History recognizes the drive you have attached to your computer and can use it to store backup copies of your files. If you don't have a drive connected, you'll see the option to select a drive. Remember — File History has to use an external drive for it to work.

TIP

File History stores information on an external drive or a network location. After you turn on File History, it will automatically create copies of your documents and photos on the drive you identify. If you lose, delete, or change your mind about changes to a file, you can restore the backup copy created by File History following the steps in the "Restore Files with File History" section, later in this chapter.

3. By default, File History backs up every personal file you have on your computer. However, you can exclude folders from this backup process in order to save space on the external drive. Select Exclude Folders on the left side of the File History window to tell Windows 11 which folders to exclude, as shown in **Figure 16-9.** Just click the Add button to add excluded folders to the list.

4. By default, Windows 11 saves copies of files every hour, but you can change this and other file history settings. In the File History window, select Advanced Settings, which you can change as follows (see **Figure 16-10**):

    - **Save Copies of Files:** This option controls how frequently File History checks for new or changed files and saves those changes. We recommend that you select Every 10 Minutes to minimize the chances of losing new or changed documents.

When you go on vacation, consider disconnecting the external drive and storing it in a fireproof safe or a safe deposit box. The drive is your insurance against theft or destruction.

- **Keep Saved Versions:** By default, File History keeps copies of your files forever. This option allows you to limit how long copies are kept. Leave this set as Forever.

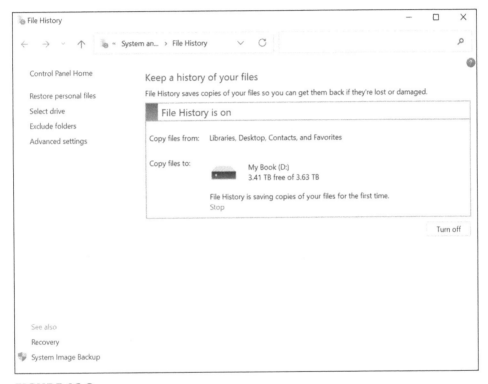

FIGURE 16-8

To turn off File History or start backing up to another drive, return to the File History window (refer to **Figure 16-8**) and use the options on the left side of the window. You can select a different drive, exclude folders, configure advanced settings, or even turn off File History.

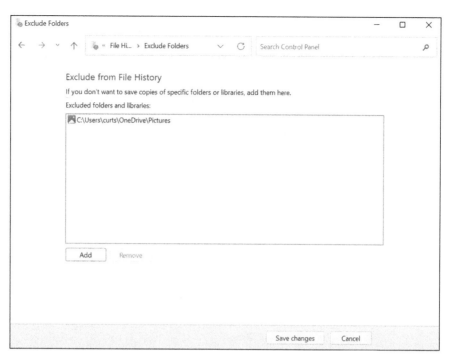

FIGURE 16-9

FIGURE 16-10

Restore Files with File History

1. Select the Search icon on the taskbar and type **file history**. On the search results screen, select Restore Your Files with File History – Control Panel. File History opens on the desktop.

2. The most recent backup versions created by File History appear in the window, shown in **Figure 16-11.** To see other versions of backups, select the left-pointing arrow at the bottom of the window. To return to the most recent backup, select the right-pointing arrow.

TIP

Don't select the Restore button until you select the specific file(s) you want. Otherwise, all files will be restored simultaneously.

TIP

Generally, you want to restore the most recent version of a file. However, if you want to restore content that you changed prior to the most recent version, browse to an earlier backup.

3. If you know the location of the file you want to restore, you can open that file's folder in the File History Restore window with a double-click or double-tap (see **Figure 16-11**). For example, if you're trying to restore a photo and you know it's in the Pictures folder, you can open that folder and find the photo you want. If you're not sure of the folder location, select the Search All box in the upper-right corner and type the file's name. Matching results appear as you type. Select the file you want to restore.

4. Select the Restore button. If you restored a file you previously deleted, you can close File History. Skip the remaining steps.

5. If the Replace or Skip Files window opens in the preceding step, a different version of the file exists in the original location. If you're sure you want to restore the previous version of the file, you can choose Replace the Files in the Destination. However, to see additional options, select Compare Info for Both Files. The File Conflicts window appears, as shown in **Figure 16-12.**

6. In the File Conflicts window, consider the following selections:

   • Select files on the left to replace files in the destination with the backup files. (This is the same as selecting Replace on the Replace or Skip Files screen.)

- Select files on the right to cancel restoring those files. (This is the same as selecting Skip on the Replace or Skip Files screen.)

- If you select the same files on both sides of the window, File History will leave the original as is and restore the backup version with the same name plus *(2)*, allowing you to have both versions. (You need this option only if you're uncertain about which version you want.)

TIP

If you're restoring multiple files at the same time, you can select different options for each: replace one, skip another, and have File History create a copy for another. That's a lot of choices in one little window.

7. Select Continue, and Windows 11 completes the operation based on your choices in Step 6. The location of the restored files opens in File Explorer.

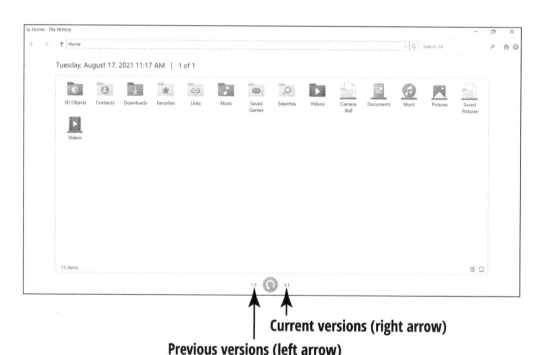

Current versions (right arrow)

Previous versions (left arrow)

FIGURE 16-11

FIGURE 16-12

Reset a Misbehaving Computer

1. Glitch happens. The computer misbehaves, a program crashes, or the machine becomes unexpectedly slow. If your computer is misbehaving, try resetting it. Select the Search icon on the taskbar and type **settings**. Select the Settings app in the search results. Under System, scroll down and select Recovery (see **Figure 16-13**).

TIP

Before you reset your PC, see Chapter 13 for information about updating and maintaining Windows 11. Updating Windows may resolve some problems.

TIP

The Reset function should leave your data alone and unchanged. However, consider following Steps 1–4 in the section "Restore Files with File History" to confirm that your external drive contains all your files. Better safe than sorry.

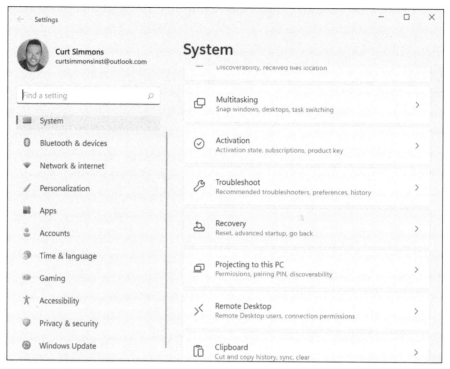

FIGURE 16-13

2. Under Reset This PC, select the Reset PC button, as shown in **Figure 16-14.**

TIP

Before choosing the reset option, try the Fix problems Without Resetting Your PC option first (refer to **Figure 16-14**). This action may fix the PC's problems much faster.

TIP

The Reset function doesn't remove apps installed through the Microsoft Store but does remove any apps you installed any other way. This safety feature is based on the assumption that something you installed from some other source is causing a problem. Be certain that you either don't need a desktop app or that you have the materials necessary to reinstall a desktop app, such as Microsoft Office. Windows 11 will create a file on the desktop after the fact, identifying the programs it removed.

3. In the Choose an Option dialog box, select Keep My Files. (Or select Cancel, if you're just exploring this feature and don't want to continue with it.)

4. Reset runs, and your computer will restart at least once. When the reset process is complete, the Lock screen appears. Sign in as usual. If you see a file named *Removed Apps* on the desktop, double-click or double-tap it to open that file in your browser. The removed apps are listed. If you're aware that one of these apps created a problem, don't reinstall it.

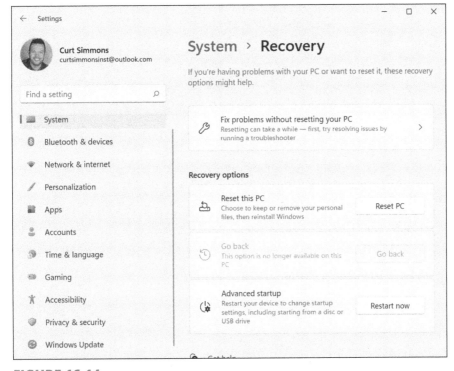

FIGURE 16-14

Don't be intimidated by the Reset function — it's easy, quick, and worthwhile if it makes a problem computer run better.

TIP

Index

About the Author

Curt Simmons is the best-selling author of nearly 100 technology books on a variety of topics. He is also a courseware developer focusing on Microsoft products and photography. When Curt is not writing or teaching, he spends his time fishing, gardening, and continuously restoring his 125-year-old Victorian home. He lives in Texas with his wife, Dawn, and two Sphynx cats.

Dedication

For Dawn

Author's Acknowledgments

I would like to thank Steven Hayes at Wiley for giving me the opportunity to write this book and Susan Christophersen for editing it. Thanks as well go to Ryan Williams, the technical editor, who made sure all instructions were accurate.

Publisher's Acknowledgments

Executive Editor: Steven Hayes

Project Manager and Copy Editor:
Susan Christophersen

Technical Editor: Ryan Williams

Proofreader: Debbye Butler

Production Editor: Mohammed Zafar Ali

Cover Image: © Ranta Images/Shutterstock

Dummies is the global leader in the reference category and one of the most trusted and highly regarded brands in the world. No longer just focused on books, customers now have access to the dummies content they need in the format they want. Together we'll craft a solution that engages your customers, stands out from the competition, and helps you meet your goals.

Advertising & Sponsorships

Connect with an engaged audience on a powerful multimedia site, and position your message alongside expert how-to content. Dummies.com is a one-stop shop for free, online information and know-how curated by a team of experts.

- Targeted ads
- Video
- Email Marketing
- Microsites
- Sweepstakes sponsorship

20 MILLION PAGE VIEWS **EVERY SINGLE MONTH**

15 MILLION UNIQUE VISITORS PER MONTH

43% OF ALL VISITORS ACCESS THE SITE **VIA THEIR MOBILE DEVICES**

700,000 NEWSLETTE SUBSCRIPTION **TO THE INBOXES OF** *300,000* UNIQUE INDIVIDUALS EVERY WEEK

# of dummies

Custom Publishing

Reach a global audience in any language by creating a solution that will differentiate you from competitors, amplify your message, and encourage customers to make a buying decision.

- Apps
- Books
- eBooks
- Video
- Audio
- Webinars

Brand Licensing & Content

Leverage the strength of the world's most popular reference brand to reach new audiences and channels of distribution.

For more information, visit **dummies.com/biz**

PERSONAL ENRICHMENT

Staying Sharp
9781119187790
USA $26.00
CAN $31.99
UK £19.99

Facebook
9781119179030
USA $21.99
CAN $25.99
UK £16.99

Guitar
9781119293354
USA $24.99
CAN $29.99
UK £17.99

Investing
9781119293347
USA $22.99
CAN $27.99
UK £16.99

Beekeeping
9781119310068
USA $22.99
CAN $27.99
UK £16.99

Digital Photography
9781119235606
USA $24.99
CAN $29.99
UK £17.99

Meditation
9781119251163
USA $24.99
CAN $29.99
UK £17.99

Pregnancy
9781119235491
USA $26.99
CAN $31.99
UK £19.99

Samsung Galaxy S7
9781119279952
USA $24.99
CAN $29.99
UK £17.99

iPhone
9781119283133
USA $24.99
CAN $29.99
UK £17.99

Crocheting
9781119287117
USA $24.99
CAN $29.99
UK £16.99

Nutrition
9781119130246
USA $22.99
CAN $27.99
UK £16.99

PROFESSIONAL DEVELOPMENT

Windows 10
9781119311041
USA $24.99
CAN $29.99
UK £17.99

AutoCAD
9781119255796
USA $39.99
CAN $47.99
UK £27.99

Excel 2016
9781119293439
USA $26.99
CAN $31.99
UK £19.99

QuickBooks 2017
9781119281467
USA $26.99
CAN $31.99
UK £19.99

macOS Sierra
9781119280651
USA $29.99
CAN $35.99
UK £21.99

LinkedIn
9781119251132
USA $24.99
CAN $29.99
UK £17.99

Windows 10
9781119310563
USA $34.00
CAN $41.99
UK £24.99

SharePoint 2016
9781119181705
USA $29.99
CAN $35.99
UK £21.99

Fundamental Analysis
9781119263593
USA $26.99
CAN $31.99
UK £19.99

Networking
9781119257769
USA $29.99
CAN $35.99
UK £21.99

Office 2016
9781119293477
USA $26.99
CAN $31.99
UK £19.99

Office 365
9781119265313
USA $24.99
CAN $29.99
UK £17.99

Salesforce.com
9781119239314
USA $29.99
CAN $35.99
UK £21.99

Coding
9781119293323
USA $29.99
CAN $35.99
UK £21.99

dummies.com

dummies
A Wiley Brand

Learning Made Easy

ACADEMIC

9781119293576
USA $19.99
CAN $23.99
UK £15.99

9781119293637
USA $19.99
CAN $23.99
UK £15.99

9781119293491
USA $19.99
CAN $23.99
UK £15.99

9781119293460
USA $19.99
CAN $23.99
UK £15.99

9781119293590
USA $19.99
CAN $23.99
UK £15.99

9781119215844
USA $26.99
CAN $31.99
UK £19.99

9781119293378
USA $22.99
CAN $27.99
UK £16.99

9781119293521
USA $19.99
CAN $23.99
UK £15.99

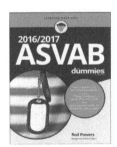

9781119239178
USA $18.99
CAN $22.99
UK £14.99

9781119263883
USA $26.99
CAN $31.99
UK £19.99

Available Everywhere Books Are Sold

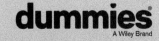

Small books for big imaginations

GETTING STARTED WITH Coding
Get Creative with Code!

Camille McCue, PhD

9781119177173
USA $9.99
CAN $9.99
UK £8.99

MODDING Minecraft
Build Your Own Minecraft Mods!

Sarah Guthals, PhD
Stephen Foster, PhD
Lindsay Handley, PhD

9781119177272
USA $9.99
CAN $9.99
UK £8.99

MAKING YouTube VIDEOS
Star in Your Own Video!

Nick Willoughby

9781119177241
USA $9.99
CAN $9.99
UK £8.99

DESIGNING Digital Games
Create Games with Scratch!

Derek Breen

9781119177210
USA $9.99
CAN $9.99
UK £8.99

GETTING STARTED WITH Raspberry Pi
Program Your Raspberry Pi!

Richard Wentk

9781119262657
USA $9.99
CAN $9.99
UK £6.99

EXPERIMENTING WITH Science
Think, Test, and Learn!

9781119291336
USA $9.99
CAN $9.99
UK £6.99

CREATING Digital Animations
Animate Stories with Scratch!

Derek Breen

9781119233527
USA $9.99
CAN $9.99
UK £6.99

GETTING STARTED WITH Engineering
Think Like an Engineer!

Camille McCue, PhD

9781119291220
USA $9.99
CAN $9.99
UK £6.99

WRITING Computer Code
Learn the Language of Computers!

Chris Minnick and Eva Holland

9781119177302
USA $9.99
CAN $9.99
UK £8.99

Unleash Their Creativity

dummies.com

Newport Community Learning & Libraries

dummies
A Wiley Brand